T0136335

Multi-Access Edge Computing in Action

Multi-Access Edge Computing in Action

Dario Sabella
Alex Reznik
Rui Frazao

CRC Press is an imprint of the
Taylor & Francis Group, an **informa** business

CRC Press
Taylor & Francis Group
6000 Broken Sound Parkway NW, Suite 300
Boca Raton, FL 33487-2742

Printed on acid-free paper

International Standard Book Number-13: 978-0-367-17394-4 (Hardback)

Visit the Taylor & Francis Web site at
http://www.taylorandfrancis.com

and the CRC Press Web site at
http://www.crcpress.com

CV 10.20.2021 2211

Contents

PART 2 MEC AND THE MARKET SCENARIOS

Foreword

I have contributed to the telecommunications industry standards for over 16 years, largely in 3GPP and ETSI, and to a lesser extent or indirectly in IETF, BBF, and in industry bodies such as NGMN and GSMA.

The authors and I first collaborated over 5 years ago at the very inception of the ETSI Industry Standardisation Group for Mobile Edge Computing, the first standards development body to produce interoperable standards and APIs for what later became multi-access edge computing (MEC) and which is still the leading body in that field. Most of us have held office in that organization. We have all contributed to its technical, marketing, or outreach publications in all that time, and continue to do so.

In its founding white paper, the ISG recognized that a new ecosystem was needed in order to serve perceived new markets that required very low communication latency or provision of content very close to its end consumer. That meant decentralizing cloud-based compute workloads and MEC was born. Communication service providers, infrastructure vendors, platform vendors, application developers, etc., all came together to produce interoperable solutions for a number of different deployment options and to raise awareness of the valuable new consumer and enterprise markets that only MEC could serve.

Five years on, much of that vision has become or is nearing reality. MEC trials are commonplace and commercial offerings exist. MEC is included in proposed solutions for 5G, factory automation, smart cities, and intelligent transport system use cases, to name just a few. Other standardization bodies and forums such as 3GPP, OpenFog Consortium, and the Open Edge Computing initiative are broadening the MEC footprint to such an extent that it has become mainstream.

This book provides the readers with a unique opportunity to learn about and assimilate the technical, deployment, market, ecosystem, and industry aspects of MEC, all in one place. Its authors are leaders in the field, and they have made key contributions in all of those areas through further specific ETSI White Papers and presentations at numerous industry-wide events.

Dr Adrian Neal
Senior Manager, Industry Standards
Vodafone Group Services Ltd.

Acknowledgments

We would like to warmly thank all colleagues and collaborators who are working with us in the area of edge computing and multi-access edge computing (MEC) standards, as in some cases the ideas expressed in this book are also a result of a collaborative effort with them. Our knowledge and expertise on this technology took advantage from the huge networking and tremendous interest of companies in this area, starting from operators, technology providers, system integrators, but also smaller companies and start-ups. The interest of the industry in edge computing is continuously increasing, so our perception of the ecosystem is taking benefit from the numerous feedback received from the various stakeholders.

On the private sphere, we would also like to thank our wives and families for their patience and support, since the writing of this book subtracted time to our family duties. So, we hope our efforts were worthwhile, and we would like to thank in advance the readers for their interest in this book.

Authors

Dario Sabella is working with INTEL as Senior Manager Standards and Research, acting also as company delegate of the 5GAA (5G Automotive Association). In his role within the Next Generation Standards division, Dario is driving new technologies and edge cloud innovation for the new communication systems, involved in ecosystem engagement and coordinating internal alignment on edge computing across SDOs and industry groups, in support of internal and external stakeholders/customers. In 2019, he has been appointed as ETSI MEC Vice-Chairman. Previously, he was serving as multi-access edge computing (MEC) Secretary and Lead of Industry Groups, and from 2015 as Vice-Chairman of ETSI MEC (Mobile Edge Computing) IEG. Prior to February 2017, he worked in TIM (Telecom Italia group), in the Wireless Access Innovation division, as responsible in various TIM research, experimental and operational activities on OFDMA technologies (WiMAX, LTE, 5G), cloud technologies (MEC), and energy efficiency (for energy saving in TIM's mobile network). From 2006, he was involved in many international projects and technological trials with TIM's subsidiary companies (ETECSA Cuba, TIM Brasil, Telecom Argentina). Since joining TIM in 2001, he has been involved in a wide range of internal and external projects (including FP7 and H2020 EU funded projects), often with leadership roles. Author of several publications (40+) and patents (20+) in the field of wireless communications, radio resource management, energy efficiency, and edge computing, Dario has also organized several international workshops and conferences.

Alex Reznik is a Hewlett Packard Enterprise (HPE) Distinguished Technologist, currently driving technical customer engagement on HPE's Telco strategic account team. In this role, he is involved in various aspects of helping a Tier 1 Telco evolve towards a flexible infrastructure capable of delivering on the full promises of 5G. Since March 2017, Alex also serves as Chair of ETSI's multi-access edge computing (MEC) ISG – the leading international standards group focused on enabling edge computing in access networks.

Prior to May 2016, Alex was a Senior Principal Engineer/Senior Director at InterDigital leading the company's research and development activities in the area of wireless internet evolution. Since joining InterDigital in 1999, he has been involved in a wide range of projects, including leadership of 3G modem ASIC architecture, design of advanced wireless security systems, coordination of standards strategy in the cognitive networks space, development of advanced IP mobility and heterogeneous access technologies, and development of new content management techniques for the mobile edge.

Alex earned his B.S.E.E. Summa Cum Laude from The Cooper Union, S.M. in Electrical Engineering and Computer Science from the Massachusetts Institute of Technology, and Ph.D. in Electrical Engineering from Princeton University. He held a Visiting Faculty appointment at WINLAB, Rutgers University, where he collaborated on research in cognitive radio, wireless security, and future mobile Internet. He served as the Vice-Chair of the Services Working Group at the Small Cells Forum. Alex is an inventor of over 150 granted U.S. patents and has been awarded numerous awards for Innovation at InterDigital.

Rui Frazao is the CTO and EVP of EMEA Operations at B-Yond. Prior to B-Yond, Rui was CTO of Vasona Networks, an innovative provider of multi-access edge computing (MEC) solutions, acquired by ZephyrTel (2018). He also previously held various group technology positions during his 15 years at Vodafone including serving as the

Director of Network Engineering overseeing network activities across Germany, the Netherlands, Hungary, and the Czech Republic. His work with Vodafone included implementing the industry's earliest VoLTE deployments and launched the first virtualized network core platforms in Europe. Rui previously held roles with Cisco, payment network SIBS, and the Lisbon Stock Exchange. He has completed studies spanning business strategies, computer systems, electrical engineering, and telecommunications.

Introduction

Edge presence is viewed as absolutely necessary to enable the key use cases driving the need for the 5th Generation of Mobile Technology ("5G"). Among these are Tactile Internet, Interactive Gaming, Virtual Reality, and Industrial Internet. All these require extremely low latency for some application components. As a consequence, physical limitations (i.e., speed of light) prohibit the execution of these components in the traditional "deep" cloud. Another set of use cases that is likely to heavily rely on edge computing is "massive" Internet of Things (IoT) – that is, IoT where a large number of devices such as sensors send a large amount of data upstream.

Market consensus is that pre-filtering of this data at the edge is necessary in order to make the overall system scalable without overloading the network and deep cloud compute resources. This makes edge presence generically critical for the success of 5G, as well as the related MEC standard. Moreover, as a recent white paper from ETSI demonstrates,[1] MEC can play an even more critical role in the industry's move toward 5G. It enables deployment of 5G applications over the existing 4G infrastructure, thus smoothing the investment curve required to get to 5G and allowing Telco operators to better align expenditures associated with the deployment of 5G with actual 5G-related revenue streams.

Making 5G a reality involves a highly diverse stakeholder ecosystem that not only includes Telco operators, vendors, application, and content providers, but also start-ups and developer communities, as well as local government and other public entities. To many of these, enabling MEC implies a significant investment in developing and deploying new infrastructure, re-architecting existing applications and integrating these with MEC services, understanding the positive impacts that a large number of micro data centers can have on local economies as well as the planning needed to allow these to spread. All these diverse entities are looking for a comprehensive understanding of the benefits of MEC from their perspective and guidance on how to solve the challenges related to 5G.

This book provides a complete and strategic overview of MEC. It covers network and technology aspects, describes the market scenarios from the point of view of different stakeholders, and analyzes deployment aspects and actions to engage the ecosystem.

As the earlier discussion makes clear, MEC exists in and supports a highly complex "5G world" in which technologists and non-technology decision makers must act in concert and do so within a large interconnected ecosystem of which MEC is just one, albeit important, part.

The book is divided into three sections, with several chapters in each, to address these three key aspects – a technology-focused section, a market-focused section, and an ecosystem-focused section.

Note

1 www.etsi.org/images/files/ETSIWhitePapers/etsi_wp24_MEC_deployment_in_4G_5G_FINAL.pdf

PART 1

MEC and the Network

1

From Cloud Computing to Multi-Access Edge Computing

This chapter introduces multi-access edge computing (MEC) from a network perspective, starting from the historical background of cloud computing, and then considering new trends (including open innovation, network softwarization, and convergence between Telco and information technology [IT]) that drove the evolution toward the edge.

Let us start, as the saying goes, in the beginning. What is MEC? To understand this properly, we actually need to start from the end, with the letter "C" denoting "Computing." This book is primarily about computing, which currently means it is about "the Cloud" (whatever that means – we'll get to that a bit later). Then, there is the "E" denoting "Edge." Chances are you think you know what "edge computing" is, but we are willing to bet that your definition is too narrow. If that's true, we hope that at the very least, this book helps broaden your "edge horizons." Then, there is the "M" which stands for "Multi-access." Here, the "access" is important – it is "edge computing" (whatever that means) which is somehow "connected" to an "access" – that is, a network that users and other client devices (e.g., those billions of things that the Internet of Things [IoT] is being built from) use to "access" the Internet. The "multi" in retrospect is the least important part of the acronym, a designation indicating that MEC technologies can be used in all kinds of networks (mobile, Wi-Fi, fixed access) to enable all kinds of applications.

And so, it appears, MEC is a kind of a chimera – a cloud technology that's located away from the cloud and that has something important to do with networking. It is also a chimera because – as we shall see – it comes about from the convergence of several disparate trends that have been around for some time. But this does not mean that MEC as a field is uncoordinated, disjointed, and nonfunctional – no more so than the mythical Chimera was.

As a brief digression for those readers with a theoretical bent, MEC represents a practical convergence of computing, communication, and – through its critical importance in enabling the industrial IoT – control: the holy grail of modern information sciences. And for those of you with a more philosophical bent, it is also a vibrant illustration of the efficiency and robustness of decentralized decision-making as compared to centralized "optimized" approaches.

1.1 To Edge or Not to Edge

A proper place to start seems to be the question of why one even needs edge computing in general and MEC in particular. Much has been written on the subject, but it can be summarized as follows: there are applications for which the traditional cloud-based application hosting environment simply does not work. This can happen for a number of reasons, and some of the more common of these are:

- The application is latency sensitive (or has latency-sensitive components) and therefore cannot sustain the latency associated with hosting in the traditional cloud.
- Application clients generate significant data which requires processing, and it is not economical, or, perhaps even not feasible to push all this data into the cloud.
- There are requirements to retain data locally, for example, within the enterprise network.

A big driver for edge computing is the IoT, where edge computing is commonly referred to as *fog computing*. NIST (National Institute of Standards and Technology), in its "Fog Computing: Conceptual Model" report [8], makes the following statement:

> Managing the data generated by Internet of Things (IoT) sensors and actuators is one of the biggest challenges faced when deploying an IoT system. Traditional cloud-based IoT systems

are challenged by the large scale, heterogeneity, and high latency witnessed in some cloud ecosystems. One solution is to decentralize applications, management, and data analytics into the network itself using a distributed and federated compute model.

Moreover, IoT is rapidly developing into a significant driver of edge computing revenue – as evidenced by Microsoft's edge cloud solution called "Azure IoT Edge."

However, IoT is just one of the several types of applications that require edge presence. In a white paper that has been widely influential in defining what "5G" is, the Next Generation Mobile Networks (NGMN) alliance lists eight classes of 5G applications that define 5G user experience and drive requirements on 5G mobile networks [9]. These include:

- Pervasive Video
- 50+ Mbps Everywhere
- High-Speed Train
- Sensor Networks
- Tactile Internet
- Natural Disaster
- E-Health Services
- Broadcast Services

A rough top-level analysis of these categories leads to a conclusion that most of them either require edge computing or significantly benefit from it. Indeed, we can make the following statements:

- Pervasive Video: edge computing can be used to significantly reduce backhaul/core network loading by edge caching and video processing and transcoding at the edge.
- High-Speed Train: such "high-speed" environments will almost certainly require application presence "on the train" to avoid dealing with network limitations associated with connectivity from a high-speed platform to a stationary network.
- Sensor Networks: The massive IoT problem of collecting and processing massive amounts of data, which is a primary focus of fog computing, lies in this category.

- Tactile Internet: use cases and applications in this category are known to require end-to-end latencies as low as 1 msec. In most networks, the physical limitations imposed by the speed of light make it impossible to achieve such latencies without edge computing.
- Natural Disaster: supporting these use cases requires deploying networks on a "connectivity island" (i.e., with limited/intermittent or even absent connectivity to the Internet). Thus, any applications have to run at the edge.
- Broadcast Services: these can benefit significantly when content can be present at the edge, as that would save significant network traffic. Moreover, edge-based contextualization of broadcast can improve what is made available in each particular area.

Clearly, edge computing is a key enabling technology for 5G, something that was recognized as early as the NGMN paper, which lists "Smart Edge Node" as a "Technology Building Block" and lists its use to run core network services close to the user as well as its potential use for application services such as edge caching.

However, focused as it was on mobile *networks*, what the NGMN paper missed is that because its "Smart Edge Node" is a landing zone for applications, it really needs to become a kind of "cloud node." This theme was picked up by ETSI (European Telecommunications Standards Institute) in the white paper "Mobile Edge Computing: A Key Technology Towards 5G" and in the creation of an Industry Specification Group (ISG) focused on what was called mobile edge computing (MEC) [10]. Within a few years, the group was renamed to multi-access edge computing (keeping the MEC abbreviation) to recognize the fact that its work was applicable across all types of access

> MEC thus represents a key technology and architectural concept to enable the evolution to 5G, since it helps advance the transformation of the mobile broadband network into a programmable world and contributes to satisfying the demanding requirements of 5G in terms of expected throughput, latency, scalability and automation.

networks: mobile (3GPP defined) as well as Wi-Fi, fixed access, etc. Again, why paraphrase when we can just quote:

One thing that was missed, or rather not highlighted by all this work, is that edge computing – specifically MEC – is not just a "5G technology." In fact, MEC is a critical tool in enabling operators to launch 5G applications on their existing 4G networks. This can have a significant impact on the business side of MEC – something discussed in detail in Ref. [11] and also in our discussion of the economic and business aspect of MEC in Chapter 3.

To conclude this brief introductory discussion, let us summarize the themes: edge presence is needed to make the full world of 5G worknetwork. This includes IoT, which is the focus of many initial deployments, but encompasses a much broader set of applications, use cases, and markets. MEC enables such edge presence by creating a cloud-like application landing zone within the access network – that is, as close to the client devices as possible. It is therefore a key enabler of the emerging world of computing – 5G, IoT, AR/VR, etc. This book expands on these themes and examines in some detail what they mean, the various ecosystem players, challenges and opportunities, as well as provides an overview of the key technologies involved. However, we must start by actually explaining what MEC is – or is not – and this is what we turn to next.

1.2 The Cloud Part of MEC

Recall, a page or two back, we noted that the primary letter in the "MEC" abbreviation is the last one – "C" denoting "computing", but really denoting "cloud." And so, we begin by looking at the cloud computing aspects of MEC. The Wikipedia page on "Cloud Computing"[1] defines it as follows.

> **Cloud computing** is an IT paradigm that enables ubiquitous access to shared pools of configurable system resources and higher level services that can be rapidly provisioned with minimal management effort, often over the Internet. Cloud computing relies on sharing of resources to achieve coherence and economies of scale, similar to a public utility.

As the same page notes, the term was popularized by Amazon Web Services (AWS) in the mid-2000s but can be dated to at least another decade prior. So, the cloud computing aspect of edge computing seems to be a well-known thing. Indeed, one of the main goals of edge computing is to "enable ubiquitous access to shared pool of configurable system resources and higher-level services that can be rapidly provisioned." A detail-oriented reader may wonder why the quote stops where it does, and indeed this is not accidental.

So, let us consider what is *not* requoted, notably "sharing of resources to achieve coherence and economies of scale." In fact, achieving what's behind these two short terms required significant advances which took several decades to be realized to a point where cloud computing became an economically viable business:

- Separation of physical hardware and applications through virtualization. This made it possible to migrate application workloads between different hardware platforms without requiring existence of different SW builds for each particular type of HW.
- Convergence to a few, industry-standardized "compute architectures" – primarily the Intel x86 architecture, so that the vast majority of applications that are virtualized are built with the assumption of an Intel-architecture–based processing underlying it.
- Development of high-speed Internet, which made possible transfer of large amounts of data and computation outside private enterprise networks.
- Development of the World Wide Web, which enabled name-based resource access paradigms. (It is unlikely that cloud computing would work well if our applications had to rely on IP for resource addressing, since IP addresses are naturally tied – that is, "pinned" – to particular HW.)
- Introduction, notably by AWS, of REST-API–based service management framework – using the World Wide Web transport mechanism (HTTP).
- An economic environment that made possible the deployment of massively sized data centers that could be turned into shared public clouds (again, led by AWS).

The above developments and resulting technologies made it possible for enterprises and cloud providers to put together systems which presented compute resources to applications as a homogeneous pool of uniform abstract resources (vCPUs, virtual RAM, virtual disk storage) that can be consumed as without regard as to their physical origin. To highlight how difficult this task really is, we note that even today the x86 CPU architecture remains the single most prevalent virtualized architecture. While ARM-based processors are ubiquitous across many industries, including, notably the telecom industry, the same is not true for ARM virtualization, which is significantly less used than x86-based virtualization. Furthermore, if you have a high-performance computing application that requires access to GPUs, then you are out of luck. Notwithstanding the extensive adoption of GPUs for a wide range of applications, GPU virtualization support is only now being developed by, for example, OpenStack. This lack of adoption is not due to any deep technical challenges. In the case of ARM, for example, the virtualization technology has been around for a while. This is because economies of scale are required for a cloud system to make sense, and enabling such a scale requires a timely convergence of a lot of factors, including existence of extensive ecosystems of applications and tools that can come together "at the right time" to make the cloud work.

So, is MEC a cloud technology? Absolutely. It is all about abstraction of compute (and storage) resources in exactly the same way as traditional cloud technology. Like traditional cloud technology, MEC leverages modern resource access paradigms (specifically Web-based resource access) and management framework; for example, all ETSI MEC APIs are defined to be RESTful and use HTTP as the default transport. However, it differs from a traditional cloud in a key aspect: scale. We are not implying that MEC (and more broadly, edge computing) lacks the same scale as traditional cloud computing. Rather, the nature of scale, and therefore the challenges that scaling presents, are of a different nature. To properly understand this, we need to examine the second component of MEC – the letter "E" denoting "Edge."

1.3 The Edge Part of MEC

There are important reasons why edge computing is now moving into the forefront of the conversation both in cloud computing and in Telco – and it is not because it is a hot new technological development. The origins of edge computing can be traced at least to Edge content distribution networks (CDNs), which were developed in the early to mid-2000s. A good brief summary can be found on the CloudFlare site (www.cloudflare.com/learning/cdn/glossary/edge-server/), which states the following:

> An edge server is a type of edge device that provides an entry point into a network. Other edges devices include routers and routing switches. Edge devices are often placed inside Internet exchange points (IxPs) to allow different networks to connect and share transit.

The CloudFlare site also provides a nice illustration, which is reproduced in Figure 1.1.

Here we make a key observation. The "Edge CDN" server as defined by CloudFlare – and indeed as is commonly used by others – is actually located at the farthermost point of the communication services provider (CSP) network (the Internet Service Provider/ISP being a special case of CSP) from the user. In part, this was dictated by a simple necessity – Edge CDN providers such as CloudFlare and Akamai simply could not get their devices any closer, since that would have meant leaving the Internet and placing their devices inside the CSP's proprietary networks.

This led to a natural next step, briefly explored, of developing a CSP-owned "Edge CDN" located as close to the user as possible.

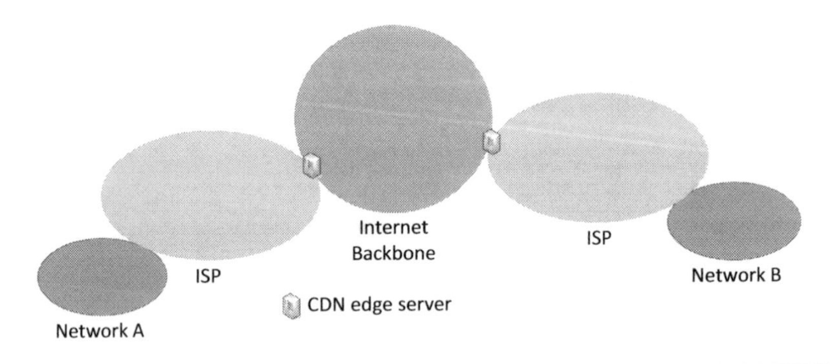

Figure 1.1 Illustration of Edge CDN. (Elaboration from CloudFlare www.cloudflare.com/learning/cdn/glossary/edge-server.)

In the case of mobile networks, this would mean locating it at the radio access network (RAN), or rather on the "S1" interface connecting the RAN to the core network. Unfortunately, doing so would require intercepting the normal flow of traffic which, in a pre-5G mobile core network, is tunneled between the RAN (e.g., the base station) and the Packet Gateway (PGW). The PGW is located at the farthest edge of the mobile network, that is, precisely where a "normal" Edge CDN is located.[2] The situation is illustrated in Figure 1.2, which shows a highly simplified diagram of the 4G core network (called the "evolved packet core" or "ePC"). The ePC interfaces are labelled, but of these, the S1 interfaces (note that there are two!) and the SGi interface are going to be really important, as they will reoccur at various points of discussion in this book as we encounter "MEC on the S1" and "MEC on the SGi" implementation options. The S1-U interface – for S1 User Plane – connects the RAN to the Serving Gateway (SGW) and carries user traffic, while the S1-MME interface connects the RAN to the mobility management entity (MME) and carries traffic which controls the various aspects by which end user devices access the RAN. It should properly be considered the "control plane." The SGi interface is the "architectural reference" the 3GPP gives to the "plain vanilla IP traffic" interface in and out of the ePC. While Edge CDNs could be placed in other logical locations in the ePC (e.g., on the S5, within the SGW), generally speaking, the assumption is that if you are going to move into the mobile network, you want to be at the RAN, that is, on the S1.

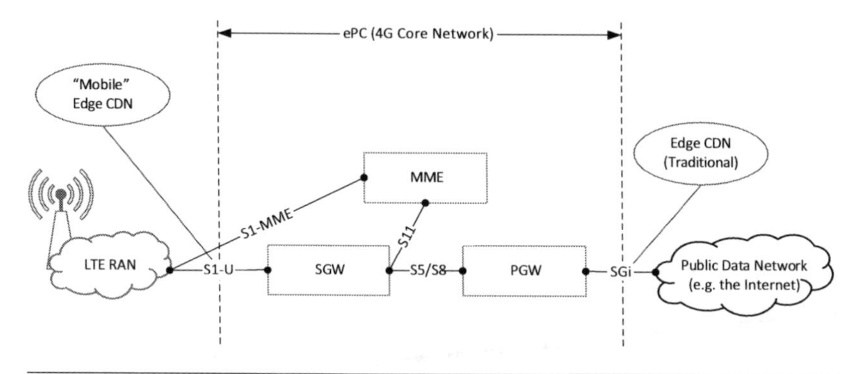

Figure 1.2 Highly simplified 4G core network showing Edge CDN locations.

The need to have an "Edge CDN" node "on the S1," as well as the potential to offer other services there, led to the development of both a standards-defined architecture (LIPA) and a "transparent" local breakout approach, in which the tunnels on the S1 are broken. Both approaches were the subject of active development and productization in the mid-2000s and, anticipating MEC, were used for applications much broader than just "edge data caching." See Ref. [3] for an example that is completely different from edge caching as well as for a good discussion of how locating processing "on the S1" works. It should be noted here that in the 5G architecture, as currently being defined by 3GPP, placement of an application function next to the RAN will be supported natively via a properly located user-plane function (UPF). We shall discuss this in more detail further ahead.

Returning, however, to the traditional Edge CDN, we note that its location at the far edge of the CSP network was dictated by an additional factor – how these entities work. At least in part they use sophisticated algorithms to analyze user population statistics and content request statistics, and attempt to predict which content is likely to be requested by which user populations. This only works if the user and content population statistics are sufficiently large and if the storage at the "edge" sites is also sufficiently large to hold a significant amount of content. Moving the Edge CDN closer to the user also reduces the size (and thus, statistical sample) of the user population served at any given time as well as the amount of content (again statistical sample) that passes the caching point. This could be counteracted by increasing the size of the cache; unfortunately, the economics of moving to the edge dictate the exact opposite – the amount of storage has to be reduced. Thus, a traditional approach to edge caching would not work much closer than the edge of the CSP network.

The way out of this is to take advantage of the rich context which being at the very far edge offers. For example, a very small cell serving a coffee shop is highly contextualized – its user population is highly likely to be into coffee and have a particular profile associated with that coffee shop. If that context can be exposed to the application, the application might just know what to do about it. If it is then provided a landing zone for its content at the small cell, it might well know what to do with this storage. This idea was recognized by the Small Cells community; the work of the Small Cells Forum in this area is documented in Refs. [4–7]. However, its realization would have to wait for edge computing to come into its own – after all, once a landing zone for content was provided for application, a natural next step is to also have a landing zone for the application's compute – at least for some components that could run at the edge.

Clearly, then, the idea of doing something at the edge is neither new nor are some of the technologies enabling it, but that still leaves open the question "what is the edge?" or, maybe, "where is the edge?" or, perhaps, "where is the boundary between the edge and the traditional cloud?" Let us now make the following bold statement. One of the more difficult questions you – the reader – may ever be asked is: "what is the 'edge' in the context of 'edge computing?'" It's easy to give some good examples. Moreover, if you are deeply immersed in your own field, you may think that *your* example is actually *the* edge. But then, talking to an acquaintance working in a related field, you find, to your surprise, that their "edge" is different from yours and that in fact defining where that "deep" cloud ends and the "edge" starts is difficult. Nevertheless, this is something we need to do – otherwise, the rest of this book is moot. We cannot very well write a book (or an essay for that matter) useful to a broad audience unless we have some common understanding with that audience of what it is that the book is about.

1.4 The Access Part of MEC

As much as we want to realize, we need to start with good examples. Amazon seems to have a clear definition of what edge is – just look at Greengrass. Microsoft more or less agrees with them, ergo Azure IoT Edge or AzureStack. This leads us to the following definition: edge is the extension of a public cloud to on-premise deployments. The idea is to provide the enterprise customer a unified management experience across both "deep" public and on-premise edge cloud and seamless automated workload migration (subject, of course, to size and other constraints).[3] In fact, in the world of enterprise computing as it stands today, this is a pretty good definition – and thus a good starting point for us.

What's missing? One instance of edge computing which is present in most enterprises and which doesn't quite fit this definition is customer premises equipment (CPE). A CPE is a component of wide-area network (WAN) infrastructure, provided by the CSP to its (usually) enterprise customers. It resides at a customer's site and terminates the WAN links into that site. Its purpose is to provide secure and reliable access between the WAN communication capabilities and the local area network (LAN) at that particular site. As such, a CPE typically comprises the switching and smart routing functionality for moving traffic between the WAN and LAN and also for load balancing across WAN links when multiple such links are present, as is often the case. Additionally, a firewall is present, providing industry standard security services to traffic passing to/from the enterprise. Both the switching/routing and firewall functionalities are typically policy-configurable and support QoS differentiation, integration with enterprise policy systems, etc. In some cases, additional functionality such as CSP-required support for charging, regulatory compliance, or Wi-Fi AP (Access Point) and LTE eNB may also be part of the CPE.

The traditional CPE consisted of discrete vertical (HW+SW) implementation of all these components. Often, but not always, these were packaged into a single "box" so that the CPE appeared as a single device to the user. Nevertheless, the internal of the CPE remained HW-based, making it inflexible and costly to upgrade, maintain, and fix.

Recently, the industry has been moving toward a flexible, configurable, and often virtualized WAN approach, usually referred to as software-defined WAN (SD-WAN). As part of this trend, there is an increasing move to replace the traditional CPE with a flexible "universal" CPE (uCPE) where all the CPE applications are virtualized and run on a generic compute platform. Figure 1.3 shows an example of a typical uCPE with a WLAN controller application in addition to the standard uCPE applications.

As is the case with other applications, virtualization brings about significant advantages: the ability to remotely monitor, maintain, and upgrade the various uCPE applications and to do so frequently and inexpensively (what used to be an HW change now becomes an SW upgrade); the ability to deliver different CPE appliances (i.e., different max throughput, max connections, users, WAN

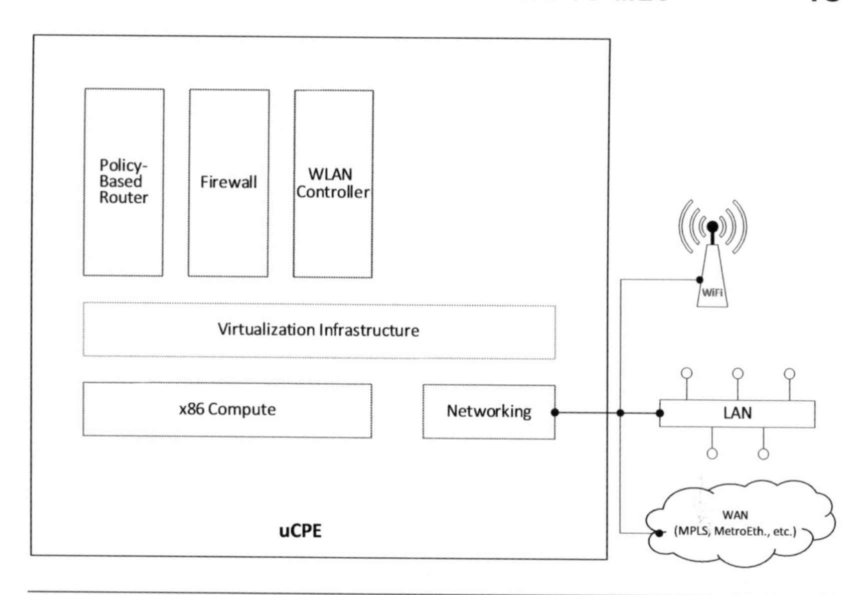

Figure 1.3 An example of a uCPE.

links, etc.) on a common platform – in fact, the ability to morph one type of CPE into another (hence, the "universal" moniker). For example, the uCPE device in Figure 1.3 may have been shipped to a customer site without the WLAN controller application. At some point, while the device is at the site, the customer requests the addition of the WLAN controller functionality. The appropriate application is pushed to the device remotely and activated, with the customer needing only to turn on and connect the WLAN AP. Provided the compute platform in the uCPE is sufficiently capable to run this additional application, all of this is done remotely, without the downtime of the WAN functions (since the other applications do not need to be taken down) and with minimal customer interaction.

At this point, it appears that uCPE is essentially yet another virtualized application (or rather a set of such applications) running on generic compute – and it should therefore be possible for uCPE to run on the same on-premise edge cloud that the enterprise runs its other applications, for example, on AWS's or Azure's edge solution. However, the applications running on uCPE differ from standard cloud applications in several critical ways.

1.4.1 Real-Time Data Processing

Operations on the real-time data stream is a key aspect of these applications – as such, they interact heavily with the "networking" component of the physical infrastructure. One common way that this is done is via data plane virtualization, that is, the Ethernet switch is virtualized on the x86 compute (Open vSwitch is an example) and the "networking" is just the physical layer function. The higher layer applications (PBR, Router, etc.) then interact with this virtualized switch. A second common approach is using SDN, in which case the "networking" component is a programmable switch (e.g., an OpenFlow switch), controlled by an SDN within the virtualization infrastructure with the higher layer applications interacting with the SDN controller.

Whatever the approach to networking is taken, the upshot of the need to operate on real-time data is that the uCPE applications are subject to very different operational requirements than traditional "IT" applications that are operated – and virtualized – by an enterprise. Application downtime – even very short downtime – can result in significant service disruption and data loss. Load balancing and resiliency through redundancy of application instances is difficult – sure, you can run multiple PBR instances on multiple compute nodes, but you cannot duplicate the data being processed by one instance at any given time.

1.4.2 SLAs and Regulatory Requirements and Critical Infrastructure

A related aspect of the uCPE applications has to do with the requirement that these are tasked to satisfy for the CSP. These include contractual service-level agreements (SLAs) – essentially performance guarantees that are associated with the connectivity services provided by the CSP to the Enterprise. Even more significantly, these communication links can often be part of critical infrastructure, that is, infrastructure where downtime may have significant and even catastrophic impact on business performance, or, in the case of public infrastructure, peoples' lives. Finally, in the case of public infrastructure, the uCPE performance is subject to various regulations –from its potential role as critical infrastructure to lawful intercept, etc. All of these cannot withstand the kind of failures that are easily addressed for "IT" applications through the now standard approaches of redundancy and load balancing.

1.4.3 Network Function Virtualization

The result of these vastly different requirements is the development of an understanding that these represent a different type of virtual applications and that the infrastructure for enabling such functions must be different as well. This different approach to virtualization is now called network function virtualization (NFV), and it is well accepted that the virtualization and management infrastructure it requires is different from the traditional virtualization world of "IT" and "Web" applications.

One key example of a management framework for such an infrastructure has been developed by ETSI's NFV ISG, and it is a framework that will be examined deeper in this book. However, as we do with the difference between "edge computing," "MEC," and a bit later "ETSI MEC," we caution the reader not to confuse the general concept of NFV with the particular management framework of ETSI NFV – the latter is both a subset of the former (with a focus on management aspects) and a specific example of a generic concept.

1.4.4 Not Your Brother's IT Cloud

So now, we can return to the question we asked some time back – why shouldn't it be possible to run uCPE applications on the same on-premise edge cloud that the enterprise runs its other applications on? The answer is quite simply because the enterprise edge clouds are just that and are not NFV clouds. They usually cannot run NFV infrastructure – and this is indeed true of, for example, the current implementation of Amazon and Microsoft Edge Environments. As a result, most NFV deployments utilize OpenStack. Moreover, a typical enterprise application does not expose the networking aspects that are required by NFV. The compute clusters (and the underlying networking) are architected to different requirements for enterprise and NFV, and thus, one should expect a different architecture to result. The NFV applications are "packaged" as virtual network functions (VNFs) which generally require significant management over and above what a standard virtualization stack (e.g., OpenStack or VMware) offers – thus the need for SW such as VNF Managers and NFV Orchestrators. Service orchestration is done using tools far more sophisticated than what Ansible or Puppet provides – not because these are bad tools – quite the contrary, their success in the enterprise world speaks wonders of how good they are. They simply do not work in the NFV space.

Clearly then, the uCPE is a different kind of beast than a typical enterprise application, but is it just an exception, a special case of something that does not have to be generalized? The answer is no, although the uCPE is probably the only example of "multi-access" edge computing that runs within an enterprise. Here are just some *locations* within a CSP network where edge clouds may be deployed – customer premises (just discussed for the enterprise, but not for residential), antenna towers, fiber aggregation points, Central Offices (COs), mobile telephone switching stations (MTSOs). The *applications* that may run in these locations include the distributed units (DUs) and centralized units (CUs) of a cloud radio access network (CRAN), packet care (ePC), border network gateways, mobile network video optimization, deep packet inspection, etc. All of these are VNFs and thus behave and have requirements akin to those of the uCPE, and differ from traditional "IT" applications in the same way that uCPE differs from traditional uCPE applications.

This does beg the question, why are there so many *locations* running so many "weird applications"? After all, it does seem like Enterprises and Web Services, as enabled by, for example, Amazon and Microsoft, are making do with just two *locations* – the cloud and the premise. Why is Telco so different? A big part of the answer is that in the traditional IT/Web services world, there are only two entities: *the cloud* and *the enterprise's premises*. The rest is just a pipe. Since the emergence of the Internet, the players in this "over-the-top" (OTT) space have never had to worry about how the pipe worked. It just did. This was – and remains – the magic of the Internet.

However, that "pipe" is actually a highly sophisticated global engineered system (perhaps the most sophisticated such system created by mankind) in which multiple highly sophisticated components are used to ensure that both the critical communication and a request for a YouTube video receive the expected QoS. And once we start talking about virtualization of the components that make up this infrastructure, we can no longer ignore its complexity – it is exposed to us, and we have to deal with it.

Moreover, this globally engineered system relies on a massively distributed infrastructure – perhaps the only distributed infrastructure of that scale in the world. Hence, it is "edge-native" – perhaps as much as 90% of Telco systems is edge. As these components are virtualized and migrated to a generic compute platform, each becomes a cloud point-of-presence. And so, voila, we have so many options.

This raises another question: does a CSP really need all of these options? For any single CSP, chances are the answer is no. Each provider is likely to pick a few edge cloud "locales" – perhaps as few as one. The choice is driven by the specifics of each CSPs: the architecture of their network – RAN, access, etc.; the kind of population demographics they are serving; and the use cases needed to enable and the business cases associated with these. All of these can vary tremendously from one operator to another and so will their definition of "edge." Moreover, they also change over time – hence the need for architectural malleability in each operator's edge architecture. Given all this variability across the CSPs in our industry, we as "the Telco industry" do need to enable all of these various options. This means that we need to develop infrastructure, standards, management frameworks, etc. that can address all of these options in a simple, unified, and *highly scalable* way.

1.5 Who Needs Standards Anyway?

Here it is worthwhile to make a brief digression and discuss the role of standardization. If you are a "Telco person," your reaction is probably "why?", "of course we need standards!" However, if you are a "cloud person," your reaction is "why?", "I haven't had the need to bother with standards so far." And so, when it comes to MEC, we once again have a set of divergent opinions on a topic of potentially key importance, which is why we need this digression.

To understand the underlying cause of this difference, we need to, once again, consider how traditional "IT" and "Web" applications are developed. The development team makes a number of key decisions: architectural approach (e.g., microservices), development and operations philosophy (e.g., DevOps), compute platform (most likely x86 if you plan to virtualize), and so on. Among these is the cloud provider/stack. Some common choices may be AWS, Azure, Kubernetes,

Mesosphere, OpenStack, etc.[4] Each one of these comes with its own approach to management services, which means, its own APIs, which the configuration, management and assurance scripts and services will have to utilize. However, this is not a problem; after all, the development team is going to pick just one, maybe two. Chances are the team is already familiar with the environment, but even if it's completely new, after some learning curve, you are up and running. Moreover, if you go with a widely used platform, there are plenty of good tools out there to help, both in the Open Source and in commercial SW space.

Let's translate this experience to MEC, keeping in mind that MEC is about the CSP provider's edge cloud – that is, the CSP becomes a cloud provider. Just for simplicity, let's restrict our attention to the United States. At the time of writing of this book, the United States has four major mobile carriers: Verizon, AT&T, Sprint, and T-Mobile. It also has several major independent broadband/cable providers: Comcast, Spectrum, Time Warner, and CenturyLink. With MEC, each one of these becomes an edge cloud provider, which appears to be similar to the AWS, Azure, Kubernetes, OpenStack, etc. ecosystem listed earlier, except for one critical point: *you can't just select one or two. As an application developer, you have to be able to work with all of these.*

The reasons become apparent if you think about your users: they have to be able to reach your cloud instances. However, while it is reasonable to expect that most of your users are able to reach AWS most of the time and also are able to reach your private cloud running OpenStack all the time, it is not reasonable to expect that customers of Operator 1 (Op1) are able to reach Operator 2's (Op2) edge cloud. And even if they are, Op2's edge cloud is NOT an edge cloud for Op1 customers. To reach it, their communication has to "leave" Op1's network, traverse the Internet, and then enter Op2's cloud. By this point, any edge benefits (proximity, low latency, minimization of network BW, etc.) are lost – in fact, the Op1 customers would be better off accessing a public cloud–hosted instance. We have illustrated this in Figure 1.4.

This creates a need for the application developers to be able to work with most of the CSPs in any geography where the application needs to be able to be present at the edge. However, for all but the largest application developers, such scaling is simply not feasible – and even

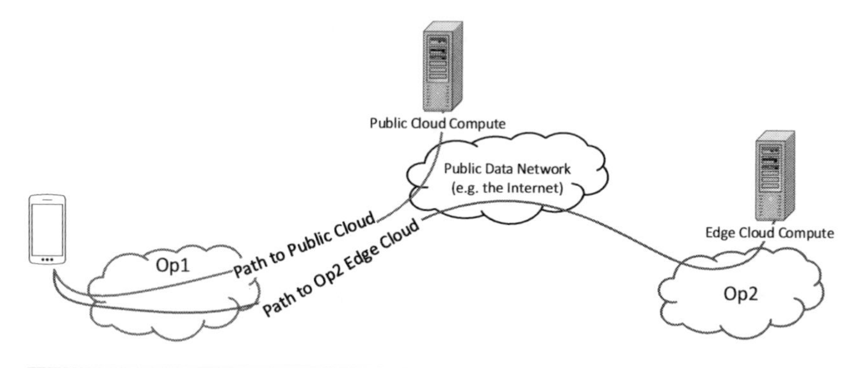

Figure 1.4 Illustrating paths to public and another operator's edge clouds.

where it is feasible, the economics involved are unfavorable. This is a problem that we will return to several times in this book, and so it will be useful to give it a short name – let's call it the Application Developers' Scaling Problem (ADSP). The ADSP has many aspects that need attention – how to set up the appropriate business relationships, specifying where to deploy applications, management of application instances, etc. One of the particular interests in this discussion is the technical problem – how can an application developer write its SW *once* and be assured that it will properly work *at every edge cloud.*

Fortunately, this problem is but a special case of a well-known problem in the communication industry – that of multi-vendor interoperability. Think, for example, of the various types of Wi-Fi devices made by various manufacturers and coming in various shapes and sizes working with Wi-Fi Access Points, also made by different manufacturers and coming in completely different shapes and sizes (from a home Wi-Fi router to Enterprise WLAN to Soft APs increasingly present in printers, etc.). And standards are the means used by the communication industry to address such a problem. When successful (and many standards, like many technologies are not), standards can enable tremendous growth of a new ecosystem, driving completely new applications and businesses. Witness the success of the IEEE 802.11 standard, which underlies Wi-Fi, or the 3GPP set of standards that have been the foundation of the global mobile communication industry since the days of GSM.

What is needed, therefore, to address the technical aspect of ADSP is a standard – more precisely, a standardized interface between the application and the MEC cloud. As we shall see further on ETSI's

MEC, standards define just such an interface. Moreover, ADSP is not the only aspect of MEC where multi-vendor interoperability is an issue, as we shall see in some of the subsequent chapters.

1.6 All We Need Is Open Source?

There is a sense in our community that while industry standard interface definitions are important, these do not have to come from traditional "standards." Instead, they can come from a development community, for example, from an Open Source project. Moreover, having these come from an Open Source project is better because the result is not just a document, but running code.

Let us agree on one point here: open source has had and is having a tremendous impact. Open source dramatically lowers barriers to entry into a market and in doing so enables small, nimble, and highly innovative companies to play on more equal footing with large established players. It does this by providing a foundation on which to build on, allowing a small team (or a large one) to focus only on those areas where they can provide value add. A dramatic example: TensorFlow allows developers to deploy a machine learning algorithm with somewhere around 20 lines of fairly straightforward Python code. This means that data companies no longer need to spend time and resources on developing the mechanics of neural network processing; rather, they can focus on where their value add is – data science.

Still, none of this means that open source serves the role that standards serve. In fact, *it does not!* To illustrate this point, let us pick on OpenStack. Not because it's bad – quite the opposite, because it is so good and so successful of an open source project, it is a good place to make a point. As anyone who has worked with OpenStack knows, No it does not – we develop to the API, but not the API itself.

- Which version of OpenStack you are developing and which APIs you need.
- Which OpenStack you are developing (the true open source one, RedHat, Mirantis, etc.)

Yes, they are *almost* the same – but *almost* is not quite the same as *the same*. And when you are a small company, having 100 edge clouds that present *almost* the same interfaces still leaves you with the challenge of scaling to integrate with 100 different – slightly, but still different – implementations. In other words, you are still missing *a standard.*

The simple truth here is that standards cannot replace open source – they don't tell you how to build anything, but open source cannot replace standards. In an ideal world, open source projects would utilize standardized interfaces in those areas where these are needed, that is, where large-scale inter-vendor interoperability needs are expected. This would, then, deliver the benefits of both worlds to the industry.

1.7 Looking Ahead … in More Ways than One

In this introductory chapter, we've tried to do a number of things: give you, the reader, a sense of why edge is important; define what edge is; and talk about the role of the various ecosystem players, including standards and open source. This was a high-level overview – the rest of the book delves much deeper into these topics and more – and it would have been both impossible and silly to try and write the whole book in the Introduction.

We also hope to have provided enough of a historical perspective to give you a sense that edge computing, in general, and MEC, specifically, is not a radical new idea out of the blue. Rather, it is a synthesis of several existing and long developing strands which came together to address an emerging need. This coming together in a timely manner is not fortuitous. It is driven by that very same timeliness, that is, by the maturity of technology to make edge computing economically feasible combined with an emergence of potential market needs.

Having said that, it would be wrong to allow you to walk away from the Introduction thinking that the major issues around edge computing have already been solved – that is far from the truth. As we shall see in some of the forthcoming chapters, what has been largely solved is the technical problem of how to stand up a single MEC site – that is, a single edge cloud at an access network, with some form of brake-out of access network traffic and some simple applications running on it. In other words, we – the broad Telco and cloud computing community – know how to build a proof-of-concept and a field trial.

Alas, once we go to production, we will have to manage thousands of small clouds. No one really knows how to do that. The Googles of the world manage tens of large clouds, not thousands of small ones. In many cases, these clouds will run workloads from entities that do not trust each other – in the formal definition of "trust," with some of these entities running SW applications that constitute critical, real-time regulated infrastructure – and others running stuff like games. How does that work? There are some potential answers, but no "best practices" – since there have been no "practices" at scale to select some "best" ones.

When these clouds do form components of critical infrastructure – or maybe just "important" infrastructure – how do they fail? What do we need to do to localize failures? Recall that IoT is a major use case for edge computing, which means edge computing will be pervasive in our lives. As many recent examples show, complex systems fail in complex ways that we do not understand and do so catastrophically more often than we think (see, e.g., Nicholas Taleb's discussion in books such as *Antifragile* [13]). Massively distributed computing makes these systems much more complex. If any of you, our readers, are looking for a challenging and impactful Ph.D. topic, this must be it.

Beyond failures, we also need to look at security – but more than just traditional cloud security. While we worry about access to our data in the "cloud" in the cyber-sense, that is, someone gaining access through user credentials theft, obtaining admin access into the management system, etc., with edge computing, we also need to worry about the physical access – quite literally, an unauthorized access to one of the many cloud sites. Prevention – while critical – can only go so far. With such a scale and most sites located in intrinsically less secure locations than a cloud provider's data center, this will happen. How can this be detected, how is sensitive data protected, and what are the recovery mechanisms?

At the same time, edge computing can be a tremendous asset in a security system. For example, Anycast IP routing is well known to be an effective mitigation strategy against Denial-of-Service (DoS) attacks – particularly, distributed DoS attacks. However, the effectiveness of Anycast-based DDoS mitigation depends on there being a highly distributed infrastructure – not just at the end points (presumably, a cloud provider can provide many end points on many compute nodes in a data center), but along the path to the end points as well. This is hard for physical reasons – eventually, the traffic must converge on one or a few physical sites where the cloud is. Not so with edge computing – the inherent scale and physical distribution of a large MEC network is a natural fit to an Anycast-based DDoS defense.

Then, there is the challenge of writing applications that take advantage of the edge. We know they should probably rely on RESTful microservices, but really not much more. From very practical approaches (see, e.g., Ref. [12]) to fundamental questions of the underlying power of distributed computation [14] (i.e., whether and how it can fully realize everything that a Turing machine can), the challenge of distributed cloud computing has not yet received extensive study – although the extensive existing work on distributed computing is bound to be relevant and perhaps the edge is where networking/computing paradigms such as ICN will find their true application. It is not an accident that Amazon named its serverless compute "Lambda functions" and that it is precisely "Lambda functions" that AWS Greengrass enables at the edge today.

The societal impact of the edge is not understood at all. From such mundane topics as business cases and strategies – which we shall address in some more details – to deeper issues of how pervasive edge cloud could impact society, very little has been studied and understood. However, it is clear that by putting flexible general-purpose computing at the edge and enabling the appropriate communication and management with it, MEC can have significant impact. From bringing cloud to the underserved areas of the world, to initiatives such as Sigfox's "Seconds to Save Lives" (https://sigfoxfoundation.org/seismic-alert/), MEC – often combined with IoT – can reshape our lives in ways that we cannot imagine today.

This brings us to the last point of this discussion – while some uses of MEC are going to happen because the society needs them, most need to be driven by solid business consideration. In short, all parties need to understand how they are going to make money. This is especially true because MEC needs to be huge so as to deliver on its promise. A dozen MEC sites in a downtown of a city is a pilot deployment, not a commercially viable enterprise. Such scale demands a huge amount of investment – and huge investment is unlikely to happen without an understanding of how a return on such investment is made. And even these business aspects of MEC are now well understood today, something which we will also explore in more depth further in this book.

Notes

1. https://en.wikipedia.org/wiki/Cloud_computing
2. Refer to, e.g., Refs. [1,2] for an excellent background on 4G networks.
3. This is the clear and commonly understood goal, even if the solutions presently offered are somewhat short of it.
4. Our mixing of VM (Virtual Machine) and container-based environments is intentional – for this discussion, it doesn't matter.

2

INTRODUCING MEC

Edge Computing in the Network

In this chapter we introduce the ETSI multi-access edge computing (MEC) standard, and the related components and main technology enablers (also by discussing the alignment with ETSI network function virtualization (NFV) standards). Moreover, we provide an overview of the MEC services scenarios, in order to offer the reader a comprehensive set of suitable use cases enabled by MEC.

So, what is MEC really? What does it look like and how does it really work? In our introductory discussion, we argued that it is a bit more than just parking and a bit of compute and storage somewhere next to a network and plugging them into the same Ethernet switch. It's time now to dive a bit deeper into what MEC really is. We start with the ETSI MEC Reference Framework and Reference Architecture, defined in Ref. [15].

The Reference Framework, shown in Figure 2.1, is an excellent place to start as it shows the essential components that make up a generic MEC system. We note here that the referenced document, being a rev. 1 document, uses "Mobile Edge ...." Terminology for MEC. Updated versions of the specifications are expected to switch to the more generic "Multi-access Edge..." terminology. To avoid confusion, we will simply use the abbreviation "ME," which can refer to both.

Starting at the bottom, we note the "Networks" level, in which ETSI defines three types of networks:

- 3GPP network. This is any type of access network defined by 3GPP. Its highlight is due to a number of factors, including the relative importance of 3GPP-defined networks in the

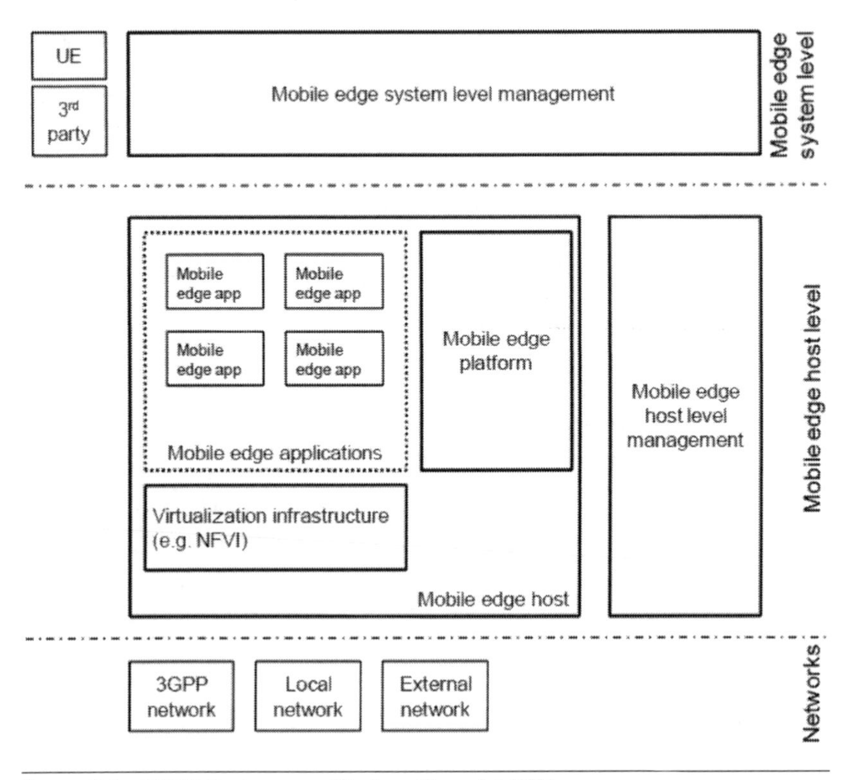

Figure 2.1 ETSI MEC reference architecture. (Figure 5-1 [15].)

mobile networking space and the consequential fact that these access networks were an initial focus of the work of ETSI.

- Local network. This refers to any other type of access network that is operated by the communication service provider. For example, provider-managed Wi-Fi access or fixed broadband access.
- External network. This is a network operated by a third party and is generally a placeholder for an enterprise network.

The "networks" space is adjacent to the "ME host" level that contains the entities that would typically reside in an edge location. This includes the virtualization infrastructure, that is, the pool of compute, storage, and networking resources as well as the capabilities to abstract and virtualize these (e.g., a Hypervisor for compute, OpenFlow switch for networking). Consuming this virtualized infrastructure is an ME platform (MEP) and a number of applications. There is a management space that contains a number of entities.

Finally, the "ME system" level contains some system-level management entities that may be accessed by a third-party application system and/or directly from a client device (UE). The entities at this level are likely to be centralized.

To further understand how an MEC system is put together, we turn to the reference architecture, also defined in Ref. [15] and shown in Figure 2.2.

An immediate observation from Figure 2.2 is that the "network level" is gone. This reflects the fact that neither the operation of the networks themselves nor the interface to ETSI MEC–defined entities is within the scope of ETSI. The goal of a "reference architecture" in a standard is to provide a reference to how the various components of the standard come together; in the case of ETSI MEC., Because network-level entities are out of scope for ETSI MEC, it is unnecessary to show these entities in the reference architecture.

On the other hand, both the ME host level and the ME system level are greatly expanded upon. While the primary reason for ETSI to do so is to properly connect the various standards they define in a bigger picture, for us this is useful as a reference MEC system design and we shall use it as such.

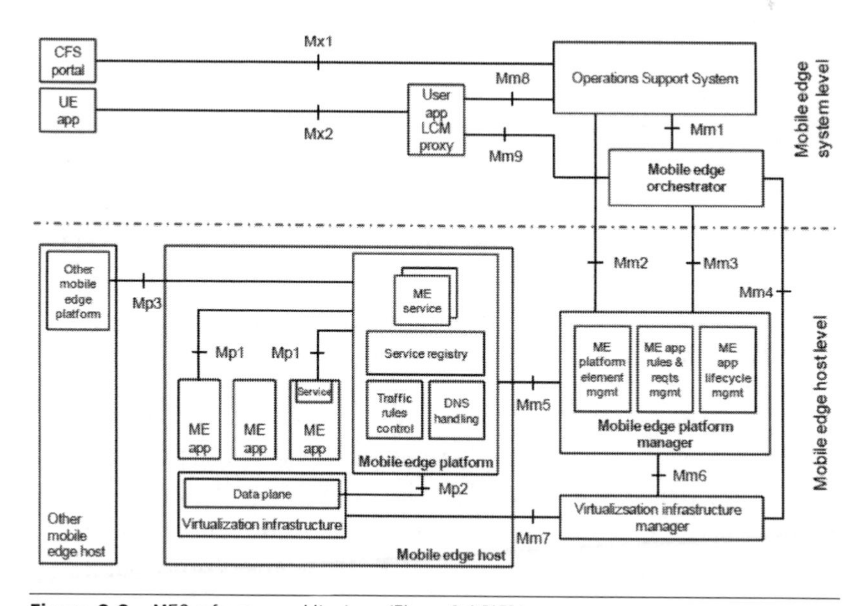

Figure 2.2 MEC reference architecture. (Figure 6-1 [15].)

2.1 The ME Host: Where the Magic Happens

Let's start, as we did with the reference framework, at the bottom and work our way to the top. We see the ME host, which is now shown in greater detail, and this *functional* component would, in most implementations, correspond to an MEC-enabled edge cloud site. As such it contains the virtualization infrastructure (i.e., generic compute and storage), which also includes the "data plane," that is, a generic networking abstraction, such as OpenFlow. With regard to the latter, we note a reference point (i.e., a set of interfaces) from the ME platform to the "data plane" called Mp2. This is the set of interfaces by which the ME platform controls the data plane. This description and our use of OpenFlow as an example of a data plane may lead some reader to infer that the ME platform should contain an SDN controller. In fact, this is just *a* potential implementation approach – one of several possibilities. The data plane/MEP may also contain some kind of a non-SDN implementation of a 3GPP 5G user plane function (UPF) and another option. The ETSI MEC set of standards does not define this implementation aspect. Moreover, given the number of available options for Mp2 (OpenFlow Southbound interface, 3GPP, etc.), ETSI MEC does not specify Mp2 at all, but enables the support of any reasonable approach.

Looking further into the ME Platform, we note a number of functionalities that ETSI MEC does assume to be present there, notably "traffic rules control" function, "DNS handling" function, and a "service registry" function. It also indicated that other "ME services" may be present. These denote some key services that the ETSI MEC *requires* to be offered by an ETSI MEC platform to any application running on an ETSI MEC host. While ETSI MEC does not define *how* these are to be implemented, it does define *what* these services are in complete detail. These definitions can be found in the specification MEC 011 [16] and include the following required services:

- ME application management. This is a basic set of services that allows the startup and termination of an ME application.
- Service registry. This set of services allows applications to find out what services are available locally at a host, how to connect to them (e.g., what the URI end points are). It also allows applications to register services they may offer.

- Traffic rule management (which a keen reader should be able to connect directly to the management of the "data plane").
- DNS rule management, which would typically be used to resolve URI and FQDN information to information that can be used to configure traffic rules using the traffic rules management service.
- Transport information query service. While RESTful services over HTTP transport are a default transport method in an ME system and ME platforms are required to support them, ME platforms may offer other transport options (e.g., messaging queues) for applications and services to use. This service allows applications and services to learn what transport methods are available.
- Time of day service. This service is used to obtain and synchronize with the timing of ME platform.
- All of these services are exposed over the Mp1 reference point, and hence MEC 011 is a specification of that reference point.

Last, but certainly not least, an ME host contains ME applications. Three are shown in Figure 2.2 because these may, in fact, come in three flavors.

The first, and the simplest, type of an ME application is best described as "just a cloud app." It consumes the virtual infrastructure resources (mainly compute and storage), but it does not utilize any of the services present on the platform. To such an application, an ME host is just another cloud site. Use cases resulting in such applications are known; the one mentioned in the Introduction uses edge clouds as a means to distribute a cloud application so as to combat DDoS attacks. However, such applications are expected to be rare. In most cases, applications (or rather application components) are located at the edge in part because they need access to the traffic on the colocated edge network. However, an ME application should not have direct access to the "data plane" portion of the virtual infrastructure; the data plane should only be accessible through the ME platform – through the Mp1, from an applications point of view.

This brings us to the second type of application and one that is likely to be the most common. This is an application that is aware of the services on the Mp1 and takes advantage of them. At a minimum, such an application is expected to take advantage of the DNS rules management and traffic rules management services. It may or may not need to use the other services defined in MEC 011.

A third type of application is one that defines its own service for other applications to use. It would then use the service registry offered by the Mp1 to offer its service and configure access parameters for other applications. This allows service vendors to offer extensive value-added services and operators to offer a rich menu of service offerings. ETSI MEC has especially recognized that while it would be unreasonable to standardize all such potential services, it is beneficial to standardize the following:

- A common set of rules for how to expose these services. This is defined in MEC 009 [17].
- A small set of service APIs that ETSI MEC believes to be of wide demand. These are standardized as optional ME services that ME Platform implementation may (but are not required to) offer on Mp1. Currently, four such optional services are defined by ETSI MEC:
    - Radio network information service (RNIS). This is standardized in MEC 012 [18] and provides operators capabilities to expose a rich set of information about their 3GPP-defined network.
    - Location API. Standardized in MEC 013 [19] and heavily based on earlier work by OMA [20,21] and Small Cells Forum [22,23], this provides operators an ability to offer rich and contextual location information, including "zonal presence concepts."
    - UE Identity API, defined in MEC 014 [24], provides a small but critically important API for identity resolution in enterprise environments (see, e.g., Ref. [25] for details).
    - A BW management API, defined in MEC 015 [26], allows operators to offer differentiated treatment to application traffic.

Additional services are currently undergoing standardization process within ETSI MEC, and the group is likely to continue to address industry needs through ongoing service API standardization.

2.2 The Magician's Toolbox: MEC Management

Let's now turn our attention to how the ME host is managed. To understand the overall management philosophy of ETSI MEC, we need to recall that a typical MEC system is likely to include hundreds or thousands of edge sites (i.e., ME host sites). These are connected to a centralized entity and/or interconnected by some type of wide area network (WAN) links. From a cloud environment point of view, these links are typically quite bad. Compared to a typical data center interconnect, these are throughput limited (too slow), latency limited (take too long), and not reliable (too prone to failure). As such, a simple translation of a data center cloud management framework to an ME system is likely to fail.

ETSI MEC recognizes this issue and enables a management framework that splits the management functions between a decentralized component that is present at every ME host and a centralized component present in the ME system. Notably, this is done in such a way as to make integration with key cloud management frameworks, for example, ETSI NFV, as straightforward as possible. However, let's put this topic aside for now and just focus on what the management framework actually does – or rather what services are defined by ETSI MEC so as to enable implementation of an efficient functional management framework.

Let's start by focusing on what is present in the host. Figure 2.2 shows the presence of a virtual infrastructure manager (VIM). As in NFV, this is assumed to be a fairly classical virtualization stack, but with a caveat. An "out of the box" OpenStack or VMware stack may not quite work in an MEC environment. As these are designed for managing a large number of compute nodes in data centers, they may often take up a fairly large number of compute resources (not an issue when the total pool is large) and assume LAN-type (high-throughput/low-latency/low-fault) interconnect between all compute nodes, storage volumes, etc. These assumptions may make it impractical to locate a full VIM stack within each edge site – the resource overhead for such a stack may be too costly. At the same time, these stacks simply do not operate well over WAN-type links that have relatively lower throughput, higher latency, and high-fault likelihoods than data center LAN links. The various developers of these stacks (both open source and commercial) are recognizing this limitation and hence are developing new VIM architectures that will allow the location of much of the VIM outside the ME host in a centralized location. Examples of such current work include the OpenStack work on this problem; see, for example, their white paper on the need for this work [27] and the work by VMware on the subject [28].

It is of importance that the movement of some portion of the VIM into the ME system level would make the ETSI MEC reference architecture inaccurate in this respect. This would potentially impact APIs defined over the Mm4, Mm6, and Mm7 reference points. However, ETSI MEC does not in fact define APIs over those interfaces as they are invariably defined by the producer of the VIM.

This leaves the ME platform manager (MEPM) as the key MEC-defined management entity in the host. This entity is responsible for life cycle management of the MEP as well as all applications running on host. This includes all monitoring, fault, and performance management. It also includes application/enforcement of policies related to application access to services, traffic flows, and cloud resources. All of these operations should be coordinated centrally, which makes the services offered by the MEPM over the Mm2 and Mm3 reference points critically important for effective management. These services are fully defined in specifications MEC 010-1 [29] for management of the MEP itself and MEC 010-2 [30] for management of the applications running on the host.

The Mm2 and Mm3 reference points allow the MEPM to offer services to the two key centralized management entities in the MEC system space: the ME Orchestrator and the operations support system (OSS).

The ME orchestrator, as described in Ref. [15], is responsible for the following functions:

- maintaining an overall view of the mobile edge system based on deployed mobile edge hosts, available resources, available mobile edge services, and topology;
- on-boarding of application packages, including checking the integrity and authenticity of the packages;
- validating application rules and requirements, and if necessary, adjusting them to comply with operator policies, keeping

a record of on-boarded packages, and preparing the virtualization infrastructure manager(s) to handle the applications;
- selecting appropriate mobile edge host(s) for application instantiation based on constraints, such as latency, available resources, and available services;
- triggering application instantiation and termination;
- triggering application relocation as needed when supported.

The OSS is a key management entity in operator networks that is responsible for a large of number of functions. As related to MEC, however, the OSS is responsible for processing and approving requests for application instantiation, termination, or relocation. These requests may be provided via the customer-facing system (CFS) portal, which is the interface to the cloud application provider, and/or from the mobile device (UE).

Enabling a UE to request application instantiation/termination directly to the OSS presents a number of potential security issues. In fact, this would be a completely new portal to the OSS – a client device has never before been enabled to interact with the OSS – but there has never before been a necessity. However, with edge computing, there is a real need for a client device to request instantiation of an application instance on a *particular* edge host, and thus the interaction is necessary. The ETSI MEC reference architecture anticipates this need by defining a user application proxy functionality whose purpose is essentially to act as a security gateway between the client devices and the OSS. ETSI MEC then standardizes an API from a UE into this proxy, which is specified in MEC 016 [31].

2.3 ETSI MEC and ETSI NFV

Although all ETSI MEC specifications are defined with the intent of enabling a self-contained MEC cloud that is able to exist in different cloud environments, the dominance of NFV in the Telco space makes it extremely important to make sure that the MEC-compliant system can be deployed and can function in an NFV-based system. To ensure this, ETSI MEC conducted a study on the coexistence of ETSI MEC and ETSI NFV, the result of which is available as a GR MEC 017 [32]. The study did identify a number of small issues, for which resolutions are being developed in ETSI MEC and ETSI NFV at the time of publication of this book. More importantly, however, the study did conclude and illustrate that the overall integration approach should be straightforward and workable even without all the identified issues being fully resolved. This approach is summarized as follows:

- Both the MEP and ME applications are VNFs.
- The MEPM acts as an NFV element manager (EM) for the MEP and for any applications that do not have their own EM.
- NFV management aspects are provided by an ETSI NFV–compliant VNFM. Notably, in practical implementation, many MEP/MEPM vendors choose to include VNFM functionality capability in MEPM, should this be needed.
- The MEP VNF needs to be instantiated first in every MEC host. As such, NFVO needs to be aware of this requirement.
- The application descriptor package (AppD), as defined in MEC 10-2 [30], is an extension to a VNFD, which is then made available to the EM (i.e., MEPM) for those applications.

2.4 MEC Use Cases and Service Scenarios

Although the ETSI MEC package of standards is intended to realize a generic edge cloud system capable of supporting just about any application, it is useful to understand what kind of use cases and service scenarios the drafters of the standard had in mind when drafting the various specifications referenced earlier (and other, forthcoming, specs). Here, two documents are of particular use. The first is the specification MEC-IEG 004 [33], which is dedicated to service scenarios. The second is the requirements specification, MEC 002 [34]. While the normative sections of MEC 002 contain just the formal normative requirements that other normative MEC specifications need to address, the document contains an extensive appendix listing the various use cases that were used in the standardization process to derive the requirements.

Let us start with the service scenarios description. As the name implies, these are broader and thus more useful in understanding the scope of the field that requires MEC. MEC-IEG 004 addresses the following seven classes of use cases:

- Intelligent video acceleration
- Video stream analysis
- Augmented reality
- Assistance for intensive computation
- Enterprise
- Connected vehicle
- IoT gateway

While not an exhaustive list – nor was it intended to be one – this list does provide an understanding of the scope of applications that either require or would benefit from MEC. Let's take a deeper look at these.

2.4.1 Intelligent Video Acceleration

While important in it and of itself, this service scenario highlights an important fact about edge computing. The presence at an MEC site of a very simple application component, one with a low computational footprint, can often make a huge difference to the overall application performance. This particular service scenario focuses on the fact that with media delivery transitioning to HTTP-based transport (which uses TCP at L4), we are increasingly encountering performance issues that are related to various known inefficiencies of using TCP over wireless networks. By including a throughput guidance component at edge, an application can properly adapt its video codec and TCP parameters to the conditions of a wireless network, thus improving the performance of video transmission from all perspectives (user experience and network load).

So what does actually have to be done at the edge in this case? Certainly, an edge-based video codec would do well, but critically, this is not necessary. What is necessary, as highlighted in the very first use case in MEC 002 (use A.2), is just a "throughput guidance" service. This service can perform as little as simply reading the radio network information (using the RNIS service as defined in Ref. [18]) and converting the information provided into a "guidance report," which is sent to a video coder/transcoder elsewhere in the cloud. Optionally, the throughput guidance component may also configure traffic rules to ensure appropriate handling of the application's traffic,

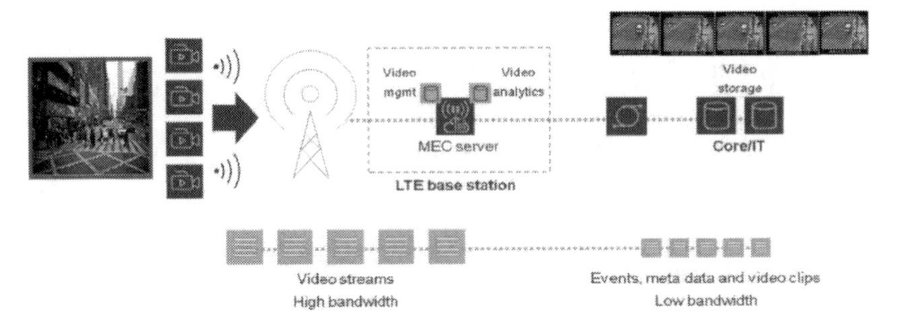

Figure 2.3 Illustration of a system implementing the video analysis service scenario. (Figure 2 [34].)

given the radio network conditions (e.g., it may decide between a mobile RAN- and a WLAN-based access).

2.4.2 *Video Stream Analysis*

If the first service scenario is focused on video *download* to a device, the second is focused on issues associated with video upload, especially in those cases where there are a lot of devices uploading such a video. Consider a surveillance system consisting of a number of cameras with a cloud-based video-analytics system. Such a system requires transmission of multiple, often high-BW, video streams for feature extraction and analysis. Moreover, the actual features of interest themselves can often be recovered from very small portions of the overall video stream. This suggests a natural edge-based preprocessing step resulting in a system that does something like the following:

- At the edge, raw video images are processed extracting only the information containing relevant features.
- The extracted information is forwarded into the cloud, which performs further image processing for extractions and then performs analysis. Figure 2.3, taken from Ref. [34], illustrates this approach.

Edge processing can once again be quite low in complexity. Consider, for example, a facial recognition system with a preprocessing component at the edge which is designed to have a low probability of missing a face, but can have a really high false-alarm probability (i.e., the probability of indicating that something is a face when it is not).

Because faces are often small components of the overall video stream, this system is still able to reduce the overall upstream data requirements by one to two orders of magnitude – a huge improvement in the overall system load. Yet, because we allow for high false-alarm probability (letting a more sophisticated final processing step correct our errors), the preprocessing step can be relatively simple and with a low computational footprint. Notably, this type of a solution is already being put into practice by the industry; see, for example, Ref. [35].

2.4.3 *Augmented Reality*

Our first two service scenarios are about managing the transmission of data between a client device and a cloud application. However, in some cases, the application itself should be local because pretty much everything it does is hyper-local. Excellent examples of this are the various augmented reality applications. For example, a smart museum augmented reality application may provide additional information when a smart phone or a viewing device (smart glasses) is pointed at a museum object, or a smart hard hat may project key information such as wiring diagrams onto components for field technicians (inspired by the Guardhat© concept: www.guardhat.com/). In all of these cases, most of the data is hyper-local – it is only relevant in the very vicinity of the physical object – the physical reality is "augmenting." As such, placing the processing and delivery of such data locally makes sense all around – from end-user QoE perspective, network performance perspective, and even system design perspective.

2.4.4 *Assistance for Intensive Computation*

From the three service scenarios mentioned earlier, one can get the impression that MEC is all about video. In fact, it is true that various operations on video are an important area of application for MEC. This is because video is often a rather extreme example of a triple threat – it needs high bandwidth requirements and low latency requirements (when real-time is involved as in some AR applications), and is streaming, which means it is very sensitive to jitter. Finally, as we noted, there is often a real *present* need for video-processing support at the edge. Nevertheless, video is still just an instance of a broader class of use cases and applications where edge presence and MEC capabilities are going to be necessary. Our next service scenario is one such case.

Consider a client device that needs to perform a complex computation that it simply does not have the computing capacity to perform. Or perhaps it does, but the battery power spent on this is prohibitive. The modern solution to this approach is to offload such a computation to the cloud, but what if one couldn't do so! Often, the rationale for even attempting computation on a client device is that offloading to the cloud is not feasible – because of latency, throughput, or some other constraint (e.g., the data needs to be kept within the boundaries of an enterprise).

By placing a pool of cloud-like compute resources very near to the client device, edge computing presents a solution to this problem. True, the computational capabilities of an edge cloud are likely to be far below those of the "deep cloud." Still, it is likely that edge clouds will be built with true cloud-grade compute nodes (i.e., real servers) and that these will be pooled. The capabilities of most edge clouds are likely to far exceed those of any client device. And so, the edge is quite literally "right there" to perform some of the heavy computational lifting. Alternately, a well-designed application will take an even more intelligent approach, identifying those computations that must stay at the edge and pushing only those into the edge cloud, while continuing to do the rest in the deep cloud.

2.4.5 MEC in the Enterprise Setting

This service scenario addresses a particular setting where MEC is expected to thrive – in fact, where MEC is already being used – that of a private enterprise. From the simplest applications like integration of IP-PBX and mobile systems (the so-called "unified communication"), to virtualization of WAN, to deployment of edge clouds in small distributed locations of an enterprise (such as retail outlets or oil pumping stations), the need for MEC in enterprise is tremendous. At the time of writing of this, enterprise is by far the largest area for commercial applications for MEC, and it includes such areas as industrial IoT (where MEC becomes the solution for large-scale IoT edge) and healthcare.

Notably, many enterprises possess extensive traditional edge clouds and have had them for some time. We are talking, of course, of the enterprise's own private cloud which is often quite close (in terms of network topology) to the end user. This begs the question of whether the enterprise even needs MEC. The answer turns out to be very much a "YES, ABSOLUTELY," and the reason MEC is needed has to do with the applications listed earlier. While a business has its own cloud (or rents a public cloud) to run its traditional business applications, such a cloud cannot support connectivity applications such as SD-WAN (discussed in the Introduction). At the same time, when the enterprise has a large number of distributed and "poorly connected" sites ("poorly connected" as compared to a data center), it often lacks the tools to deploy and manage a highly scaled edge cloud across these sites. However, the enterprise's connectivity provider has already solved the problem – it has connectivity applications running in all these sites. As such, offering edge cloud services back to the enterprise becomes a natural value proposition.

This service scenario is important for another reason. It highlights the need to integrate with traditional cloud providers like Amazon Web Services (AWS) and Microsoft's Azure. Most IT cloud applications have been written for such cloud stacks and with their recent aggressive move into the edge, this is unlikely to change. However, this is not a direct competitive threat to the connectivity provider, since the cloud providers need the connectivity provider's MEC point of presence for their edge solutions. Moreover, the primary MEC relationship for the enterprise customer is often with the connectivity provider, not with the cloud provider. Notably, solutions to address this need are emerging; see, for example, a recent effort by HPE, AWS, and Saguna [36].

2.4.6 Connected Vehicles

Another huge potential application area that is difficult to imagine without MEC is vehicular automation. The strict latency required to meet full automation (see, e.g., Ref. [37]) combined with the need to perform computation on an aggregate of multi-vehicle data makes it hard to think of a vehicular automation system design without MEC. While not as near-term as the enterprise space of applications, this area (often called V2X) is likely to become just as, if not more pervasive. Moreover, because of the integration of physical and digital worlds, public and private domains, and critical infrastructure, the challenges are immense.

At the time of writing, ETSI MEC is well underway toward addressing these challenges, having produced a detailed study of impacts of V2X on MEC [38] and initiated the required standardization activities.

2.4.7 IoT Gateway

The final service scenario listed in MEC-IEG 004 is the IoT Gateway. An IoT Gateway is a key function – usually realized in a stand-alone device – in the exploding domain of low-power IoT sensors and actuators. Essentially, it is the first "point of contact" for these devices with the rest of the world. Depending on the specific use case and the nature of the actual IoT devices, it may serve any number of purposes, including communication aggregation, computational offload, identity proxy, etc. As such, the physical realization of such a device is typically a generic compute node, integrated into an access network. This makes it an MEC Host.

2.4.7.1 Public Edge Clouds As we noted, while MEC-IEG 004 represents a good compendium of MEC services scenarios, it is by no means complete – nor was it meant to be. In fact, there are two service scenarios not in MEC-IEG004, but which are worth mentioning since these are likely to be critical to the future of MEC. The first of these is a public edge cloud, that is, the extension of a public cloud (e.g., AWS) to an MEC system. As we shall discuss at some length in the next chapter, such a scenario can play a significant part on the MEC ecosystem by enabling support for small application developers that typically develop only for one or two global public clouds.

Notably, significant progress toward enabling this service scenario is already being made. We have already mentioned the HPE/Saguna/AWS integration highlighted in Ref. [36]. Perhaps a more significant development is the trial deployments in China, of a China Unicom-based MEC system with Tencent cloud platform integrated on it. Some details on this can be obtained from their

MEC Proof-of-Concept Wiki here: https://mecwiki.etsi.org/index.php?title=PoC_12_MEC_enabled_Over-The-Top_business. By the time this book is published, the system may very well be broadly available across China.

2.4.7.2 Support of Operator Services The last service scenario mentioned here is often understood implicitly, but it is worth calling out and mentioning explicitly, because it is going to be perhaps the widest used scenario and it is also critically important to understanding how MEC is deployed in an economically feasible fashion. This is, of course, the support for operator's network services and applications. It is not unreasonable to assume that most MEC sites will be hosting some kind of operator service. This fact drives a number of key aspects in the design of MEC systems, specifically:

- The need to integrate with an NFV management framework (whether ETSI or an alternative like ONAP).
- The need to be able to support multiple "zones" (e.g., via tenant spaces) that deliver very different SLAs to their tenant applications and must be securely isolated to a very significant extent. At the very least, there is likely to be at least one "operator's zone" supporting NFV-based SLAs and critical infrastructure and at least "public cloud zone" or an "enterprise zone" in every MEC deployment.

The challenges in system design presented by this are nontrivial, although addressable. We shall dive into these deeper in this book, but also refer the reader to external white papers on the topic; for example, ETSI MEC's paper on integration with CRAN [39] (which, as we shall see is a key driver for MEC deployments).

3
The Three Dimensions of MEC

This chapter introduces three main aspects that should be considered for the success of the adoption of multi-access edge computing (MEC) technology: deployment (making it work), operation (making it work at scale), and charging/billing models for MEC (making it have a business sense).

In deciding to deploy an MEC system, we face several challenges – the first of these is, quite simply, how do I design a system that (i) meets my needs and (ii) can be deployed today. As our discussion in the first two chapters indicates, this is not a straightforward affair. No single standards body defines a complete working system (nor, as we argued, should they!). As of the time of writing of this book, there was no open source community available that provided a relatively complete design as a starting point (although a few are working toward one). In fact, as our discussion clarifies, given the highly variable nature of how communication services providers (CSPs) are likely to deploy MEC, it is not likely that any single complete open source project will meet the needs of all CSPs.

Thus, we are left with the challenge of having to put together an MEC system that is reflective of the needs of each particular operator. This chapter is about how to address this challenge, but only partly so. A major point of this chapter is that this challenge is just one of the three major challenges associated with getting a commercially viable MEC system up and running as part of a mobile network ... or rather mobile cloud. In fact, it may be the easiest of the three, because, as we shall see, we largely do know what to do and much of the technology needed is not new. The other two challenges – operating an MEC system at scale and making such a scaled MEC system profitable – are likely to be much more formidable, in part because the industry is just beginning to figure out how to address them. In this chapter, we provide guidance on how to go about thinking about these problems in your system and in your business.

Let us think about the challenges, or rather sets of challenges, as decision and design problems that lie on three dimensions: infrastructure/operations/commercial. Thinking of these as dimensions makes it clear that first and foremost, these are directions of thought when it comes to MEC.

3.1 The Infrastructure Dimension

We start with the dimension we identified first – how to design an MEC system. To make this more precise, our goal is to stand up something like a "commercial field trial" of an access network with an edge cloud colocated with that network. Because this is a field trial, the scale is going to be small and we are not going to worry about how to monetize it. But we do need it to work in a commercial network with commercial devices.

In principle, this is simple. We simply attach a small cloud to an IP router positioned at or near an access network, as shown in Figure 3.1. In this case, the MEC cloud is nothing more than a standard edge compute point of presence: there is nothing "MEC specific" about

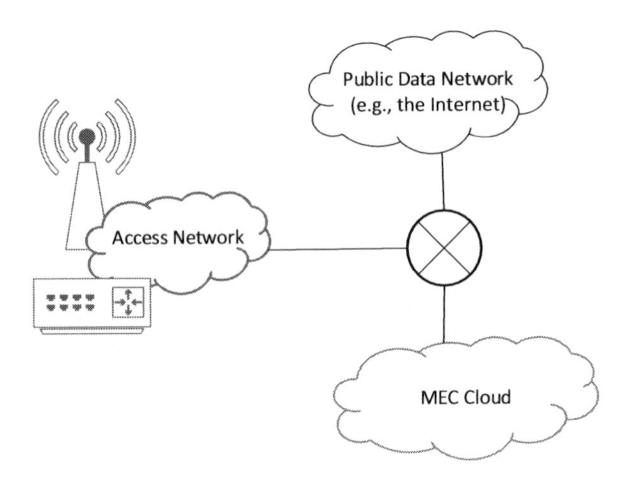

Figure 3.1 Attaching an MEC cloud to an access network – the simplest case.

it. In fact, for some "MEC" deployments, for example, in enterprise scenarios where an IT cloud is attached to existing enterprise networks and only existing IT applications need to be supported, this is sufficient. However, in this scenario, it is also not possible to expose MEC-defined services, for example, an appropriate Network Information Service, to applications in the "MEC cloud." To do so, we need an ME platform (MEP) that is set up to communicate with and control the access network as needed to provide the service. This is shown in Figure 3.2.

A more complex issue has to do with the nature of the access network that the MEC cloud is proximate to. For a typical WLAN and most fixed access networks, the traffic that leaves the access network uses publicly routable IP addresses, and therefore, the diagrams in Figures 3.8 and 3.9 suffice. However, this is not the case with mobile (3GPP-based) networks, where traffic leaving the access network is tunneled inside GTP-based bearers. We briefly alluded to these challenges in Chapter 1 (see Figure 1.2 and the text associated with it), but here we consider it in more detail. In doing so, we need to treat 4G and 5G mobile networks very differently, as the architecture of the 5G core network makes support for MEC much more natural. Our discussion is based extensively on two white papers dedicated to these topics and the references in those papers [40,41].

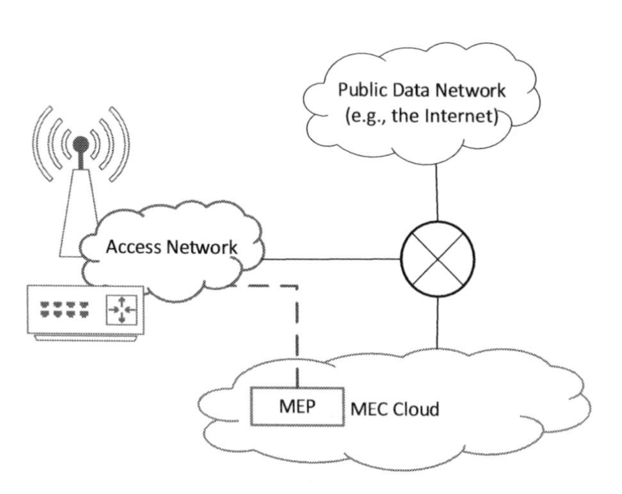

Figure 3.2 Attaching an MEC cloud to access network with MEP.

3.1.1 Enabling MEC in 4G Networks

Referring to [40], we note the basic taxonomy of various approaches to enabling MEC in a 4G network as introduced there:

- "Bump in the Wire" (BiW)
- Distributed SGW with a Local Breakout (SGW-LBO)
- Distributed S/PGW
- Distributed ePC

Referring back to Figure 1.2 in Chapter 1, these represent a progression of where MEC is located in the 4G architecture, with BiW locating an MEC cloud on the S1 (and thus also sometimes called MEC-on-S1), SGW-LBO locating an MEC cloud between the SGW and the PGW, and the last two options locating it behind the PGW. We note here that the locations we talk about here are network locations, not physical locations.

Clearly, then, we have a number of options and none of these are universally better or worse than others. The appropriate choice of option depends on the specifics of each use case, target applications, and the network operator's reality. Moreover, with the emergence of network virtualization and its enablement of network slices, an operator can support different options across different network slices. In the next section, we have summarized each option with a view to allow you – the reader – to make choices that fit your needs.

3.1.1.1 Bump in the Wire Let us start with the BiW approach. An architecture diagram depicting this approach is shown in Figure 3.3.

A major challenge here, as the name suggests, is that this approach requires the MEP to "bump," that is, to intercept the S1 interface in the 3GPP network. This interface carries GTP-bearer traffic, usually encrypted using IPSec. As such, it requires the MEP to institute an "authorized man-in-the-middle attack" on these bearers. Fortunately, techniques for doing so have been well-known for some time, usually under the name of "Local Breakout." To achieve this, the MEP needs to have access to security information (e.g., bearer key matter), as well as the ability to read and, if needed, modify control signaling on the S1-C (the control portion of the S1 interfaces). This is achieved using appropriate interactions with the core network's MME, PCRF, and

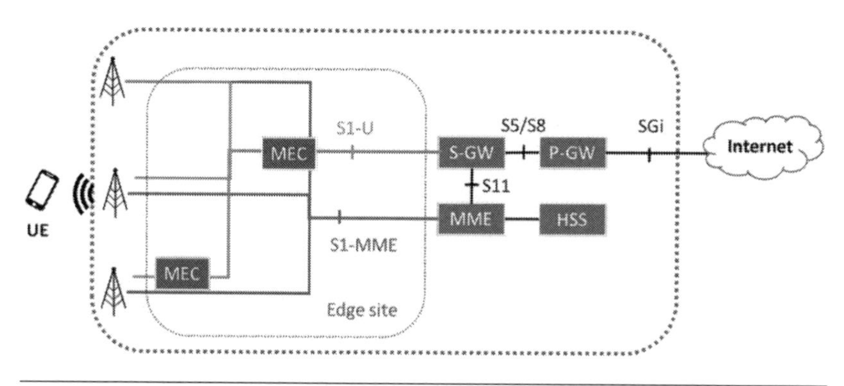

Figure 3.3 Bump-in-the-wire architecture.

other entities. Lawful intercept, charging, and other requirements can likewise be satisfied. However, we note that these operations are all outside the scope of the 4G standards, and hence, a standardized BiW solution does not exist.

Thus, a BiW is a complex solution to implement, but it comes with a number of benefits. First and foremost, by avoiding the need for its own PGW, the BiW approach has a minimal impact on the mobility of a mobile device attached to the network. This can be important, especially in public network use cases, although the concerns about connectivity to MEC applications in such cases need to be addressed. It is also the most flexible in terms of physical location of the MEC cloud, allowing colocation with RAN sites such as antenna sites and CRAN aggregation sites. We highlight this in Figure 3.3, by showing an MEC cloud (red box) located in multiple places in the edge cloud. In fact, such multistage MEC is also possible with a BiW approach.

Another benefit of the BiW approach is that it does not require core network elements to be moved into the edge site. This makes it a good approach for deployment into existing 4G networks, introducing edge computing capabilities into these networks and thus allowing introduction of 5G applications using the existing 4G networks – see Ref. [40] for more detail on this last point. We shall also return to this point when considering the business dimension of MEC.

Another benefit of the BiW approach is its flexibility in supporting different modes of MEC traffic manipulation. Reference [40] defines three such modes:

- **Breakout**. Here, the session connection is redirected to an MEC application that is either hosted locally on the MEC platform or on a remote server. Typical breakout applications include local content distribution network (CDN), gaming and media content services, and enterprise local area network (LAN).
- **In-line**. Here, the session connectivity is maintained with the original (Internet) server, while all traffic traverses the MEC application. In-line MEC applications include transparent content caching and security applications.
- **Tap**. Here, the traffic is duplicated and forwarded to the tap MEC application, for example, when deploying virtual network probes or security applications.

Because a BiW approach can be made transparent to the user plane of the rest of the 4G network, all three traffic manipulation approaches can be supported with relative ease.

3.1.1.2 Distributed SGW with Local Breakout The SGW-LBO approach, illustrated in Figure 3.4, may be a good option for an operator that does not want to tamper with S1-U tunneling, but does not want to deploy functionally distinct PGWs – and thus create separate APNs – for applications running in the edge.

This solution moves the LBO point into an SGW where the S1 bearer is naturally terminated. The SGW is then enhanced to support breakout via network-specified traffic filters. The solution retains much of the functional benefits of the BiW approach at the cost of introducing a more complex SGW into the network, requiring the presence of an SGW in the edge (thus potentially limiting how close

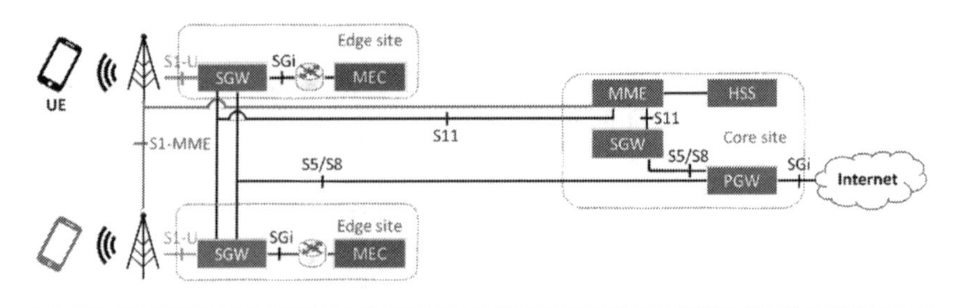

Figure 3.4 SGW-LBO architecture.

to the physical RAN locations deployment can occur). We note that this enhanced SGW is still not standardized – thus, like BiW, this approach requires implementation that goes beyond the 4G standard. Additionally, it is no longer transparent to an existing ePC; that is, its introduction is somewhat more complex.

3.1.1.3 Distributed S/PGW The next solution moves the MEC point to the outside of the PGW, as shown in Figure 3.5.

In this approach, all of the user plane functions (UPFs) of the ePC have been brought into the edge site and the MEC cloud is located behind the PGW. This brings a number of benefits. First, it is simple – from the point of view of MEC, this has the feel of a non-mobile deployment, Figure 3.2, with the MEC cloud operating on routable IP traffic. It also requires no implementations *within the 3GPP domain* that go beyond the scope of the 4G standards. Thus, conceptually, it is simpler.

However, this approach also comes with a number of limitations. First, a PGW is often associated with an APN. Here, the mobile operator must decide how this should be handled. The number of edge sites in a typical network is likely to be in the hundreds, if not thousands. An APN per site may or may not make sense. However, a single APN for all edge sites may not make sense as well. Thus, APN design (something that was probably never an issue before) may become a challenge in this case. A related complication is mobility – UEs attached to edge site PGWs cannot be mobile – inter-PGW mobility is not supported.

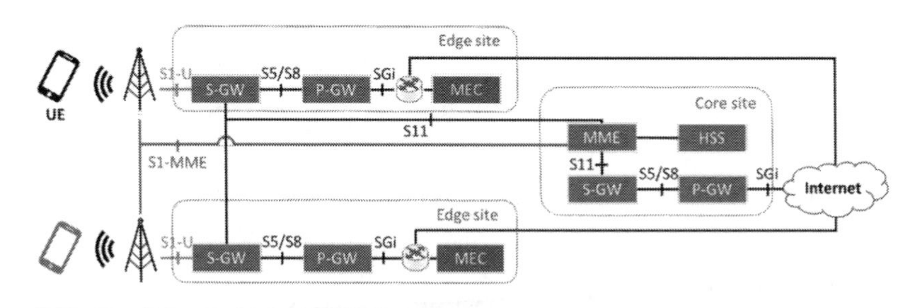

Figure 3.5 Distributed S/PGW architecture.

Nevertheless, there are use cases where the limitations of this approach are actually advantages. For example, when access to MEC applications is geographically limited to a venue or an enterprise site, lack of mobility support becomes a plus – you want the application session to drop once the UE leaves the designated coverage area. And the APN is naturally associated with the venue, enterprise, etc.

A final note before we leave the distributed S/PGW approach. It represents a great example of Control/User Plane Separation (CUPS) – something that becomes critical when we consider 5G. Note that all the control plane entities are kept at the Core site, while the user plane entities associated with edge APNs are moved to the edge. Thus CUPS, while a central design principle of the 5G core, is not purely a 5G feature. It can be done in the ePC and, as we see here, can represent the right approach to certain scenarios.

3.1.1.4 Distributed ePC One of the features of the distributed S/PGW approach we just considered is that while the user plane is fully distributed, it is controlled by a single set of control plane entities: MME, HSS, PCRF, etc. In most cases, this is, indeed, the right approach, as it allows the operator to continue managing its network as a holistic entity.

However, in some cases, it may make sense to completely distribute the entire ePC. An example is when a large enterprise customer wants a full-blow "private LTE" experience, that is, it wants the "look and feel" of owning its own small LTE-based network on its site, but using licensed spectrum. The operators (who hold the license to the spectrum) can enable this by deploying a full miniaturized ePC at the edge site, as shown in Figure 3.6.

The major aspects of this approach are similar to those of the distributed S/PGW – that is, the need for edge APNs and mobility limitation – except in this case, these are quite likely to be a desired aspect of the "private LTE" network. Additionally, as illustrated in Figure 3.6, multiple such networks can be deployed supporting different applications – and different devices. Moreover, with virtualization, this is easily accomplished as multiple network slices over the same compute, network, and RAN infrastructure.

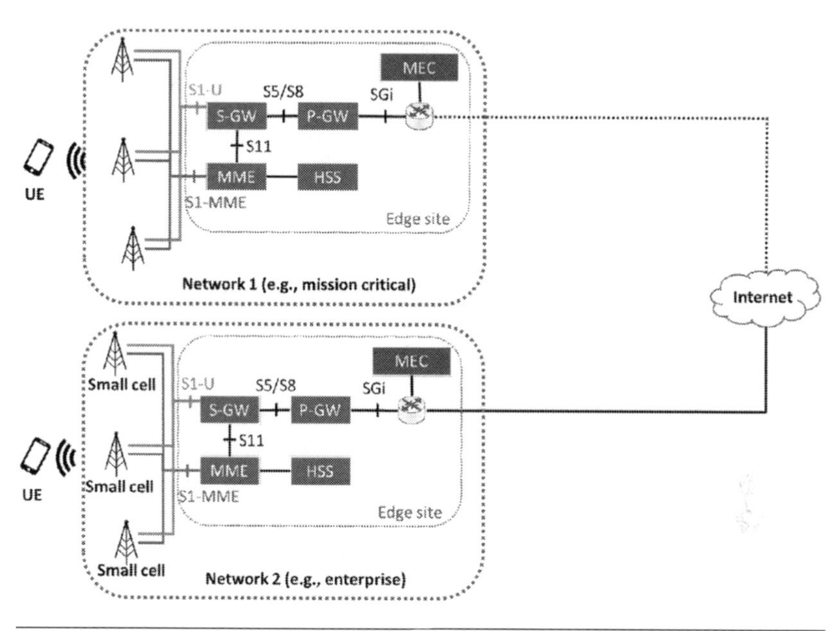

Figure 3.6 Distributed ePC architecture.

3.1.2 MEC in 5G Networks

In our discussion of integrating MEC into 4G networks, we focused on how to do this considering that the core network (ePC) was never designed to support edge computing. In contrast, in designing its 5G core network, 3GPP specifically considered edge computing needs. In particular, unlike the PGW, the 3GPP UPF can be located in multiple physical locations, and a large number of UPFs can be supported within the core network without affecting mobility. Thus, in considering MEC in 5G networks, we focus on how MEC integrates with the UPF and the service-based architecture (SBA) components of the control place part of the core network.

To be more specific, the following are some of the key enablers of edge computing in 5G as highlighted in Ref. [41], with the full list of enablers found in Ref. [42]:

- Support of local routing and traffic steering, in particular by allowing session to have multiple N6 interfaces toward the data network.

- Enabling application functions (i.e., applications external to the 5G Core Network) to influence selection and re-selection of UPFs and traffic.
- Explicit support of a local area data network, which, in most cases, corresponds to an edge cloud.

To fully appreciate the enablement of MEC by the 5G network, we start with the 5G Network architecture as defined by Ref. [42]. For the purposes of our discussion, the simplest, non-roaming, case suffices. This is shown in Figure 3.7 (reproduced from Figure 4.2.3-1 of [42]).

Let's highlight some key takeaways for our discussion from Figure 3.7. First, we need to note where the user plane and the control plane are. The user plane is effectively the bottom four entities in the figure: UE, (R)AN, UPF, and DN, although the UE and Access Network (AN; RAN when it's the 3GPP radio access network) contain control plane entities as well. DN refers to a data network – as in some external data network that is outside the core network. Thus, the only pure user plane function is the aptly named User Plane Function (UPF). This, as we noted previously, is one of the key 5G core network entities for integration with MEC.

The rest of the 5G Core network entities are all control plane functions. Notably, they are shown as sitting on a common "bus," implying a service-based approach. In fact, this is explicit, see Section 4.2.6 of Ref. [42], with the technical realization specification [43] further defining the use of HTTP as a transport mechanism and suggesting

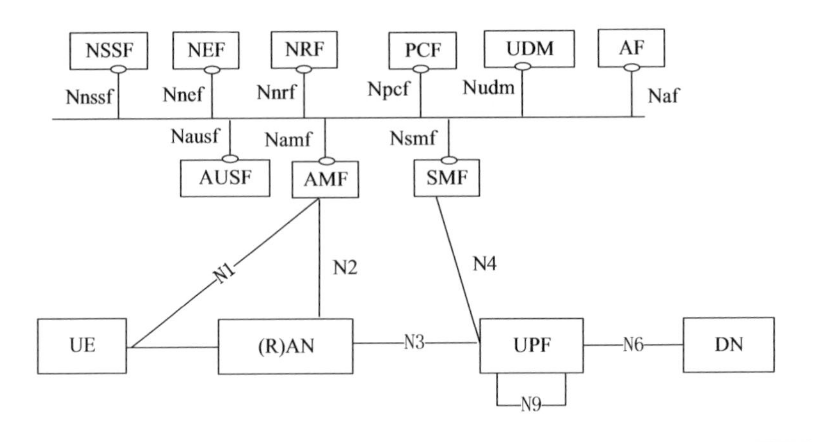

Figure 3.7 5G system architecture. (Figure 4.2.3-1, [42].)

the use of REST-based implementation by defining how the various control plane functions in Figure 3.7 can be realized in a stateless fashion (see Ref. [43], Section 6.5). Thus, for the purposes of our discussion, it may be useful to think of the control plane as implemented using RESTful HTTP-based transport. When we do so, we should remember that other transports (e.g., message queues) may also be supported by the various 5G Core implementations.

As such, the 5G network architecture solves many of the challenges of integrating an edge cloud with the mobile core and accessing internal mobile core information needed to provide MEC services. The integration problem then becomes one of how to do it properly – something that we will cover in detail in Chapter 4.

3.2 The Operations Dimension

At this point in the book, you've successfully deployed an MEC field trial in your network – that is, you've implemented one of the options discussed in Section 3.1 and your MEC cloud is now accessible. To make this deployment commercially useful you likely have to duplicate this deployment over hundreds, if not thousands of sites. In doing so, you are presented with a new set of challenges:

- Sites are different in nature requiring different types of compute. Some sites can host a relatively large number of compute nodes, while others may only host one or two. Moreover, while some sites satisfy the environmental requirements of a data center (in terms of cooling, filtering, space, power, seismic protection, etc.), others may not. Thus, while you may be able to impose the same *functional architecture* across all sites, that is, the same choice among the options in Section 3.1, it is highly unlikely that you will be able to impose the same cloud architecture across all sites.
- Sites are connected by the wide area network (WAN) links. This is important because compute nodes within a data center are interconnected using LAN links. Compared to LAN links, WAN links typically have much higher latency and are slower. WAN links are also often shared with the network provider's user traffic – thus overloading them, especially at peak times, degrades the quality of service the network

providers experience. These issues are significant because most existing tools for cloud management tacitly assume that LAN interconnects and thus may not be usable for management of a distributed edge cloud (MEC or otherwise).

- Finally, each MEC site may need to actually support multiple cloud stacks, or more specifically cloud domains. We shall refer to these as XaaS domains since they may be Infrastructure-as-a-Service domains, Platform-as-a-Service domains, or a mix of these. The need for multiple XaaS domains is being driven by a mixture of application needs and business reasons. The business reasons will be discussed in Section 3.3 in depth. Let us briefly overview the application need drivers for this source of complexity.

Think of the edge cloud that a CSP has deployed next to its access network, and in fact, let's imagine that this is a modern (5G or rather advanced 4G) network. Moreover, let's suppose that the CSP wants to utilize this edge cloud for the following purposes.

- Its own network services that need to run in the edge in order for the CSP to achieve the desired system performance. These may include:
  - Virtualized/Cloud RAN (vRAN or cRAN), which may require to operate either on bare metal or using specialized real-time–enabled virtual infrastructure.
  - Other virtualized network functions (VNFs), for example, a virtual 5G UPF, which, while running on a standard VIM, requires to be managed within an network function virtualization (NFV) framework.
- Third-party (revenue generating) applications, which may also come in different flavors, for example:
  - Applications virtualized using a traditional VM-based approach and expecting a traditional virtualization infrastructure, for example, OpenStack or VMware.
  - Applications enabled using a collection of containerized microservices and managed by Kubernetes.

Clearly, the CSP faces a challenge. It can try and manage everything under a single VIM, for example with OpenStack. In this case, the vRAN bare metal resources would be managed via Ironic, VM-based third-party applications would be managed as part of the NFV domain and containerized application would be confined to a single (presumably) large domain with Kubernetes managing resources just within this VM. While theoretically possible, such an architectural decision does come with a number of limitations. For example, limitations of Ironic; the fact that third-party applications are not VNF and are not designed to be, thus potentially leaving the CSP with the task of onboarding them into an NFV environment; performance issues associated with running containers within a VM; security and reliability implications of mixing network functions and third-party applications; etc. Whether or not the benefits outweigh the complexities is a subject for a separate discussion – our point for now is that while in some cases such a single-IaaS domain architecture works, in many cases, it may not make sense. The issue becomes even more complex if one starts thinking about supporting multiple (and competing) cloud service providers such as Amazon Web Services (AWS), Azure, and Google Cloud Platform, within a single edge site. The point is this – as a CSPs edge cloud evolves, there is significant likelihood that support for multiple XaaS domains will be required. The resulting management system architecture is illustrated in Figure 3.8, which shows a system with *K* MEC sites supporting as many as *N* different XaaS domains per site. As shown in Figure 3.8, from a CSPs point of view, this requires the presence of some kind of local sites orchestrations unit that is connected to a centralized site orchestration entity located in the CSP's private cloud. Depending on how the CSP offers the edge site infrastructure to its tenants, the local/centralized O&M may vary. For example, if the CSP offers the infrastructure as a bare metal level, then the local site O&M could be just the out-of-band infrastructure management units (e.g., HPE's iLO, Dell's Remote Access Controller) or a federation of these, with the centralized O&M being an aggregated management solution for physical infrastructure. As another example, if the edge infrastructure is being offered as NFV-based virtual IaaS, then the edge O&M becomes a combination of ETSI NFV VNFM and ETSI MEC MEPM, while the centralized O&M should include the functions of ETSI NFV NFVO and ETSI MEC MEAO.

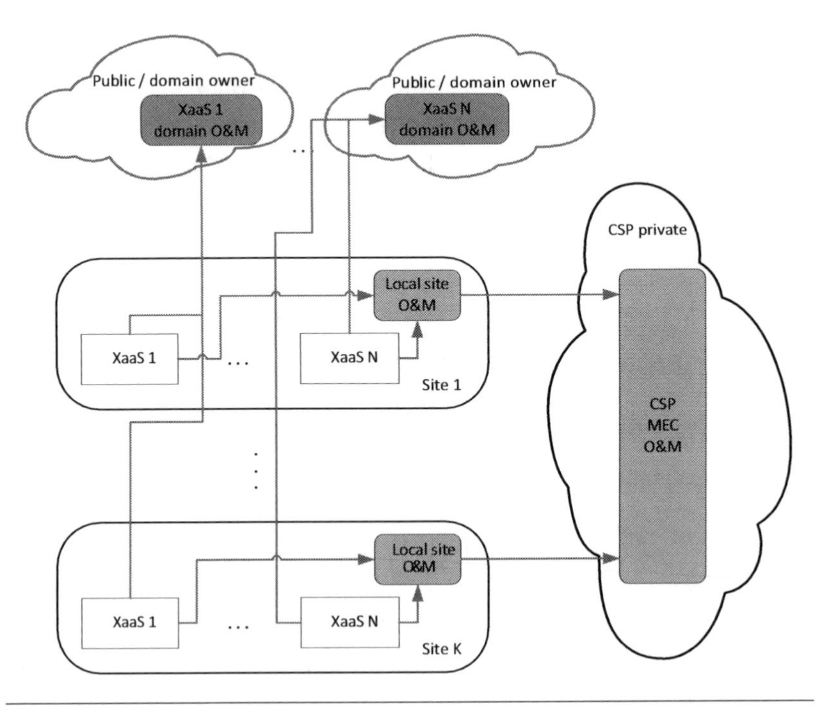

Figure 3.8 Management system architecture.

Figure 3.8 also shows the presence of XaaS domain O&M for each of the XaaS domains – these may be located in a private cloud of the XaaS domain tenant or in the public cloud. Some examples for illustration purposes are as follows: if the XaaS domain is Azure IoT Edge, then Microsoft's IoT edge management system is the O&M for this domain; if the XaaS domain is the CSP's NFV domain, then the ETSI NFV NFVO, ETSI MEC MEAO, Services Orchestration, etc. are the O&M for that domain.

At this point, the architecture presented in Figure 3.8 should raise some questions and concerns. Among these are the complexities associated with multi-domain O&M, especially making all of this work at scale and across multiple industry players; the potential need to coordinate between domain O&M systems and MEC O&M as well as potential for overlap between these – as evoked by the fact that ETSI NFV NFVO and ETSI MEC MEAO appeared in our examples as both an MEC O&M component and a domain O&M component; and the absence of FCAPS data collection and analytics in Figure 3.8 and in our discussion.

Although this may not be immediately clear, thinking about the O&M problem as in Figure 3.8 significantly simplifies the multi-domain aspects of MEC by eliminating the need for coordination across domains and between CSP and XaaS domain O&M – but still allowing for it if necessary. Again, this is best illustrated by an example. Let's assume that a CSP is running an NFV-based MEC infrastructure with KVM/OpenStack as the underlying NFVI/VIM. As part of this infrastructure, the CSP has an agreement with Microsoft to create Azure IoT Edge points of presence at each site, with Azure IoT Edge being allocated a large VM within which it can run all of its resources. The VM is allocated in an OpenStack tenant space that is outside the NFV-managed environment. In that case, the architecture in Figure 3.8 reduces to that of Figure 3.9.

Some of the key observations of this system are as follows:

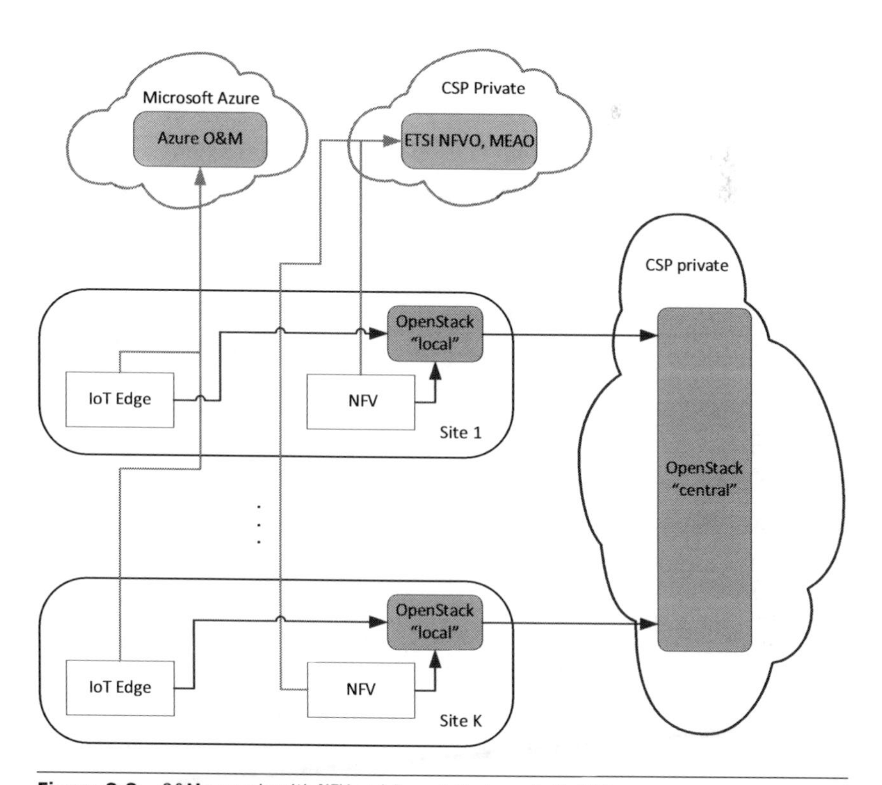

Figure 3.9 O&M example with NFV and Azure IoT edge outside NFV.

- Clear **demarcation of responsibilities** between the OpenStack O&M on the one hand and the NFV and Azure O&M on the other. By this we mean that OpenStack allocates resources to NFV and Azure O&M at a macro level – VMs. Each one of these can then decide what happens to within each VM and and how it happens. NFV O&M assembles these into VNFs, while Azure O&M further allocates them to the various services that run within its VM.
- An **as-a-Service** approach to interactions between the various O&M actors. In particular, OpenStack exposes a number of services for VM management via its well-known API, which both NFV and Azure O&M then use to change the number of resources they use (scaling vertically and/or horizontally), configure, and monitor its resources, etc.

We should also note the **lack of maturity** – at the time of writing this book – of solutions. A reader well familiar with OpenStack should note that the "local" and "central" aspects of OpenStack are not actually available – at least not at the time this book is being written – although work within the OpenStack Foundation and elsewhere is on-going.

An important observation is that the as-a-Service approach to O&M interactions requires a well-defined set of service APIs that should also be standardized in some fashion – via either a formal industry standard or a de facto one. In this case, OpenStack's API is a well-known industry de facto standard. It is worthwhile noting some of the key industry standards and when and how these may be useful. We stress that this is not meant to be an exhaustive list.

- Redfish and associated efforts (e.g., Swordfish): these are REST APIs for out-of-band management of physical compute and storage infrastructure that are seeing rapid and broad adoption.
- IPMI: an older standard for out-of-band management of physical infrastructure. Widely adopted, but likely to be replaced by the Redfish family of standards.
- OpenStack: as noted in our example here, part of OpenStack's success implies that its API has become a kind of industry standard for requesting services from a VIM.

- VMware vSphere APIs: perhaps the most significant private VIM and API.
- ETSI NFV: REST APIs both for managing NFV infrastructure.
- ETSI MEC: REST APIs for managing NFV-based services across a distributed edge (i.e., as a supplement to NFV management APIs when NFV is deployed in an MEC environment as opposed to a data center).
- Azure Stack API: industry standard APIs to be used when Azure Stack is used as the underlying virtualization infrastructure and manager.

At this point, we could also ask what happens if in our example we want everything, including the Azure IoT Edge VM, to be managed under the NFV framework, that is, the Azure IoT Edge VM becomes a single-VM VNF? Wouldn't this create a conflict between the "NFV domain O&M" and the "MEC O&M"? The answer is no, as Figure 3.10 illustrates. The NFV "domain O&M" simply

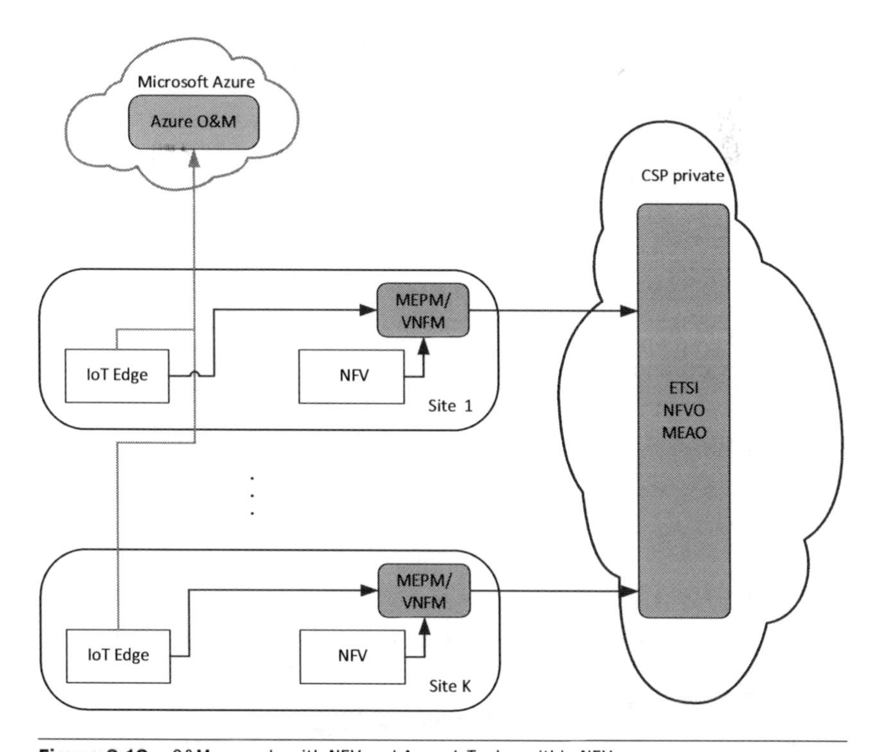

Figure 3.10 O&M example with NFV and Azure IoT edge within NFV.

disappears – it is not needed and the "MEC O&M" can handle that task. This illustrates another important point – in many cases, **domain O&M may not be necessary** and the overall system is significantly simplified.

What about the ability to scale the number of sites to hundreds or thousands or even higher? This is addressed by making sure that the interface between the "local site O&M" and "MEC O&M" is well-defined and satisfies certain requirements:

- Stateless: scaling stateless instances is significantly easier than scaling stateful ones; additionally, it makes the overall approach robust to communication session issues.
- Loose latency requirements: recall that WAN connectivity may carry high latency.
- Robust to communication session interruptions.
- Built-in and robust identity, naming, site and capability discovery, authentication, etc. system: so that the management of site identities, names, security, is taken care of.

In this respect, a standard approach to REST (i.e., REST over HTTP) represents an excellent fit. It is stateless by design and the HTTP transport is by default latency-insensitive and supports strong security (via HTTPS). It is descriptive, making site and capability discovery straightforward, and a number of well-known and robust tools can be brought to bear to support this function as well as authentication and other capabilities. These include the use of URIs for naming of resources – which can be sites as well as their capabilities; DNS for URI resolution (and thus site discovery in a network); OAuth for authentication and related functionality, etc.

It is notable that the majority of API solutions in this space are adopting REST as either the only or the default/preferred solution – this includes ETSI NFV and MEC, Redfish, and Azure Stack APIs. Well-defined RESTful APIs, in turn, allow modern O&M tooling to be defined, and although the industry is in an early stage, solutions for MEC O&M are emerging.

Lastly, let us turn to the absence of FCAPS support in Figure 3.8. This is left out mainly for clarity; otherwise, the diagram would be rather busy. The reader should assume the presence of data collection for FCAPS at all key points of the system. A more complex question is where should this data be processed and analyzed. This decision, again, is likely to be heavily driven by the design of each particular system and heavily influenced by the capabilities of modern Machine Learning approaches (including Deep Learning artificial intelligence algorithms). As such, we can assume several features:

- Learning is most likely to take place in centralized O&M locations. It requires large data sets and significant amount of processing.
- Inference is likely to be split between localized inference engines and centralized ones.
- Localized inference engines are designed to serve dual roles. First, these are used for initial data pre-filtering designed to limit the amount of data sent upstream. Second, these are used to make decisions where latency is important. In all cases, a low-complexity approach to edge inference should be used.
- Centralized inference engines are used to perform large–data set learning based on the maximal available data and often to adjust parameters in the localized inference engines and also to drive learning.
- Data should be shared between O&M systems to the extent possible – modern learning systems do best when as much data as possible is provided to them. However, in most cases, data sharing between O&M systems belonging to separate commercial entities is likely to be severely restricted by regulatory restrictions, privacy policies, and other commercial considerations.

Let us summarize our discussion on operations with some key takeaways. While the industry is in the early stage of addressing the O&M need for MEC, solutions are emerging. It is likely, however, that a CSP will need to integrate multiple partial solutions in a way that is highly specific to their own needs to achieve a complete solution. In doing so, a CSP should look for a number of key aspects: (i) ability to satisfy all the unique requirements of MEC as highlighted above; (ii) standardization at key interfaces as highlighted in Figure 3.8; (iii) appropriate implementation of these interfaces, with REST APIs being a particularly well-suited modern approach; and (iv) a well-designed modern approach to distributed machine/deep learning to support FCAPS.

3.3 The Commercial Dimension

Consider some of the design decisions discussed: the types of applications to support at the edge; whether to integrate edge offerings from major cloud providers like Amazon, Microsoft, and Google; and at what level to offer edge cloud infrastructure. To a very significant extent, the right approach needs to be driven by commercial considerations and this is the discussion we turn to in this section. A good start is to consider the macro-level business models that a CSP can use to approach the edge computing business. A good analysis of these is provided in Ref. [44], which proposes five categories of business models, with definition as reproduced below:

- **Dedicated edge hosting/colocation**: the CSP provides its customers with rack space, power, cooling, access, and backhaul network connectivity. It may also provide cloud infrastructure (this is the key point considered in Ref. [44]) or the customer may bring their own cloud infrastructure – as might be the case with, for example, AWS Outpost. Either way, the customer gets exclusive and complete control over a certain number of physical resources.
- **Edge IaaS/PaaS/NaaS**: in this case, the CSP operates as a cloud provider, providing customers distributed compute and storage capabilities, a platform for developing applications on the edge infrastructure and network services, as well as APIs and VNFs in an "as-a-Service" manner through a cloud portal as the customer interface.
- **Systems integration**: the CSP offers custom turnkey solutions for enterprise customers with specific requirements, which are (partially) met by the MEC functionality.
- **B2B2x solutions**: the CSP offers edge-enabled solutions to enterprise customers. As with existing B2B solutions, these may be for the customer's internal purposes, such as to improve existing processes, or may contribute to an end-customer offering (B2B2x). In general, these solutions will be closer to an "off-the-shelf" product than a totally bespoke offering, thus requiring significantly less integration work than SI projects.

- **End-to-end consumer retail applications**: the CSP plays high up the value chain, acting as a digital service provider for consumer applications. MEC-enabled services in this category will leverage the benefits of MEC, namely, low latency, high throughput, and context awareness, to provide consumers with innovative applications (e.g., VR for live sports).

To these, we add a sixth category – "optimization of internal operations." In this case, the edge cloud is used by the CSP to optimize its own operations (and thus reduce OPEX or improve customer QoS). Although "inwardly facing," this approach is likely to become an important component of many CSP's approaches to the cloud.

The next step is to attempt to classify these approaches to edge business. The reference [44] approaches this by looking at two ROI metrics. The first is "comparative terminal value" – essentially the potential for substantial long-term returns. The second is "three-year internal rate of return" – a measure of return on investment in near term. For the purposes of our discussion, we will simplify these into "long-term opportunity" and "short-term returns." Based on discussion in Ref. [44], these can be visualized as in Figure 3.11. The precise reasoning for positioning these as such can be found in appropriate literature, for example, Ref. [44], and may or may not apply to each particular CSP's circumstances. However, Figure 3.11 illustrates a broader point that does appear to be common across most CSPs considering how to monetize their edge deployment.

The desired approach, as represented by the red ellipse, is to find an approach that provides a reasonable immediate return on investment while also representing a significant long-term opportunity. This is especially important since deployment of an edge cloud can often require significant investment – recall that such a cloud may consist of hundreds or thousands of sites. Thus, some immediate monetization of the investment is usually required – making it as a purely long-term speculative bet is just too risky. On the other hand, without a significant long-term growth opportunity, the business case is likely to be unattractive.

Unfortunately, as Figure 3.11 makes it clear, such an approach does not exist. However, this does not mean that both goals cannot be achieved at the same time. Many of the solutions that fall into diverse

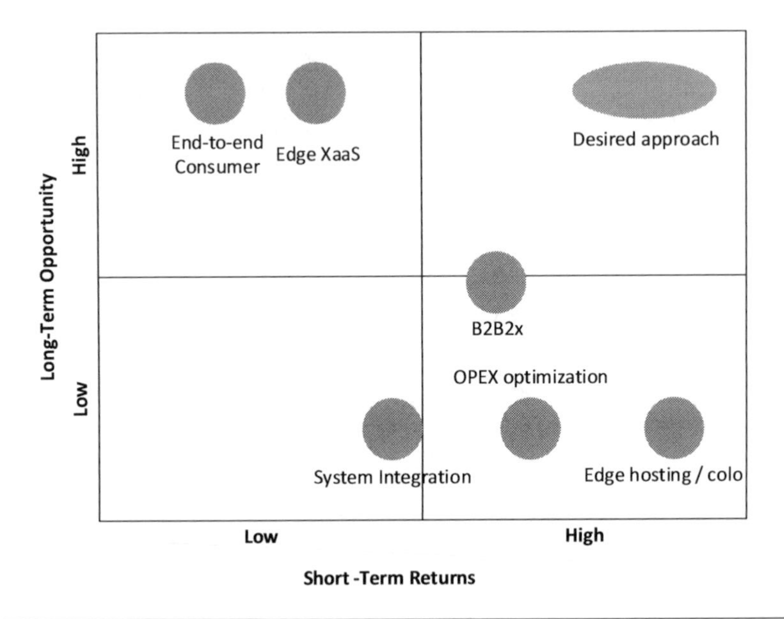

Figure 3.11 Visualizing business models [44].

business model categories are highly complementary and the flexible nature of the edge cloud allows deploying architectures that support very different business cases. The answer is therefore to think of multiple business cases that allow one to achieve the ultimate goal of a highly profitable edge cloud.

In doing so, one must be careful. Not all approaches are complementary and some early decisions may impact long-term opportunities. For example, developing a colocation relationship with a public cloud provider early on may provide the much-needed immediate revenue, but may make it harder for a CSP to develop its own XaaS edge service later. Nonetheless, in many cases, a complimentary approach is possible and rather easy to implement. For example, imagine a CSP deploying an edge cloud for the purposes of improving its internal OPEX. To accomplish this, the CSP deploys an NFV-based XaaS domain across its network. However, during the process of architecting, designing, and deploying the MEC cloud, the CSP does the following: (i) overprovision the cloud resources (compute, networking, storage); (ii) procure and/or develop operational tooling capable of supporting multiple XaaS domains as discussed in Section 3.2; and (iii) select an MEP provider capable of offering value-add MEC services (for example, those mentioned in Section 2.1) to potential future deployments. While the CSP has clearly invested more into this edge deployment than would have otherwise been necessary, the scope of the overinvestment is not huge: some extra cloud capacity; somewhat more capable tooling and MEP, which presumably cost somewhat more. Depending on the nature of VNFs supported, the overinvestment amount may actually be rather small by comparison. The biggest delta is in the planning effort – the CSP is planning for future growth opportunities without having to fully predict what they may be.

Consider what happens if the CSP is now interested in deploying IoT Edge services across its edge cloud (e.g., enable Azure IoT Edge presence across the cloud). The goal is to capture some of the revenue that cloud-based IoT services are generating. The CSP's O&M systems are ready to support a multi-XaaS environment and the infrastructure is in place for the initial launch, and the MEPs in edge clouds are ready to offer these IoT applications value-add services. The cost of what would otherwise may have been a very expensive speculative play has been minimized by bootstrapping capabilities for new long-term business growth to an NFV deployment that is generating immediate returns.

The example given earlier highlights the critical importance of being able to "think multi-XaaS" when thinking about MEC and having access to and/or developing the operational tooling capable of supporting this, as we discussed in Section 3.2.

Importantly, it also points to an approach that CSPs can take to deploy their edge presence that delivers the effective ability of the "desired landing spot" in Figure 3.11, even though no single business case can get us there. The approach is to bootstrap deploying a strategically positioned edge cloud to something that has to be done anyway and looks roughly as follows:

STEP 1: Identify the initial "reason" to deploy MEC. This "reason" is likely to be a business need, the case for which falls into the lower right quadrant in Figure 3.11. The exact nature of it does not matter, except that you should think through the limitations that your choice imposes of any future business opportunities.

STEP 2: Architect/design the edge cloud according to the following principles. (i) The application(s) associated with the business need identified in Step 1 are workloads running on a generic cloud. The nature of this particular cloud domain should be appropriate for this type of application. (ii) At least one other workload running in the same cloud domain and at least one other cloud domain of a different nature (e.g., Kubernetes vs. NFV) are to be supported at every edge site. (iii) At least one XaaS domain is to be managed by a domain O&M that is independent from your MEC O&M.

STEP 3: Develop/obtain solutions that support the design in STEP 2, paying particular attention to O&M solutions.

STEP 4: Start the initial deployment with some small overprovisioning of your MEC cloud beyond what the initial "reasons" identified in STEP 1 needs.

4
MEC and the Path toward 5G

In this chapter, we analyze all the drivers for the evolution toward the edge, from a network technology point of view, starting from key performance indicators (KPIs) to the evolution of new user devices and terminals, and also describe the consequent impact on spectrum demand (regulated at both regional and global scales), which is on its turn determining the level of success of new communication technologies.

4.1 Network Evolution toward 5G

The typical increase of data traffic demand that we all have been witnessing in these years is a consolidated trend that seems to continue also in the future. Many studies [45–48] are in fact confirming this trend, and often are updating older forecasts, as they were considered too conservative, since new estimations are instead projecting toward higher data volumes and traffic needs. Figure 4.1 shows that mobile data traffic will increase sevenfold between 2016 and 2021 globally, at a compound annual growth rate (CAGR) of 47%, with annual traffic exceeding half a zettabyte in 2021. As an additional remark, the mobile devices usage (both via cellular and Wi-Fi networks) will account for 47% of total IP traffic by 2020, a testament to the significant growth and impact of mobile devices and lifestyles on overall traffic, while fixed traffic will fall from 52% of total IP traffic in 2015 to 33% by 2020 [46]. For this reason, the evolution of mobile network is critical for the growth of the overall IP traffic.

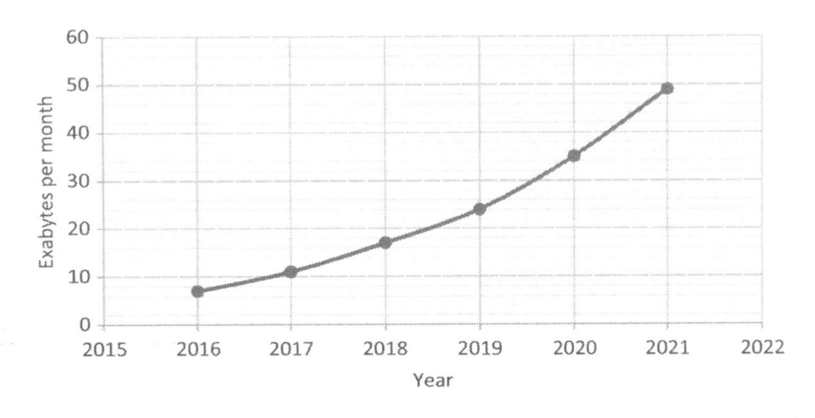

Figure 4.1 Growth of global mobile data traffic. (Elaboration from Ref. [46].)

This huge traffic demand confirms that the society is moving toward a data-driven world, and it is also a natural driver for communication network evolution (especially mobile networks). This market demand is, in fact, pushing many technical requirements to communication networks, which are obliging the entire ecosystem (network operators, technology and service providers) to continuously introduce elements of innovation in the network infrastructure and terminals, for the evolution toward future 5G systems. At the same time, it's worth noticing that the actual usage of new networks (e.g., 5G) will be influenced and determined by the introduction in the market of new terminals. The importance of new devices is discussed later on in the chapter, but at this point it is worth highlighting that this phenomenon is true since the era of smartphones, where the actual usage of new and performing devices acted as a catalyzer to stimulate the creation and consumption of new services, thus acting as a further enabler for data traffic demand (which is again driving a further cycle of network evolution).

In summary, we could imagine a sort of cycle of innovation (depicted in Figure 4.2) where the involvement of both networks and terminals are not only driven by traffic demand but they also act as drivers for the same market evolution, encouraging an increased data consumption. As quoted by Brian Krzanich (former Intel's CEO), "Data is the new oil" [49].

The typical communication network evolution, across decades, related to the adoption of the various network generations, is

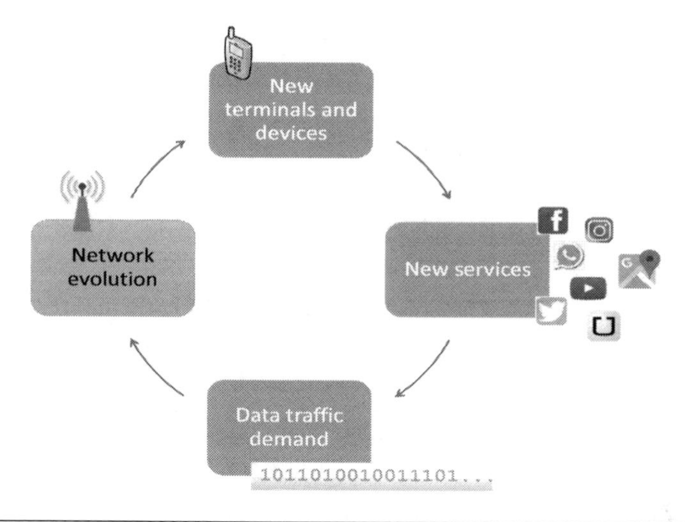

Figure 4.2 The typical innovation cycle.

characterized by the presence of waves (for each generation), which generally start after a period of research and development (R&D), consequent standardization, then experimental activities and trials for an early introduction, and gradual deployment toward a more mature market adoption and progressive consolidation (until the natural disappearance).

Figure 4.3 shows these typical waves for the past network generations (2G, 3G, 4G); anyway, as a natural consequence, a reasonable projection can be done for the future evolution toward the 5G wave: nobody can predict the future, of course, but there is a general consensus within the industry in considering some similarities

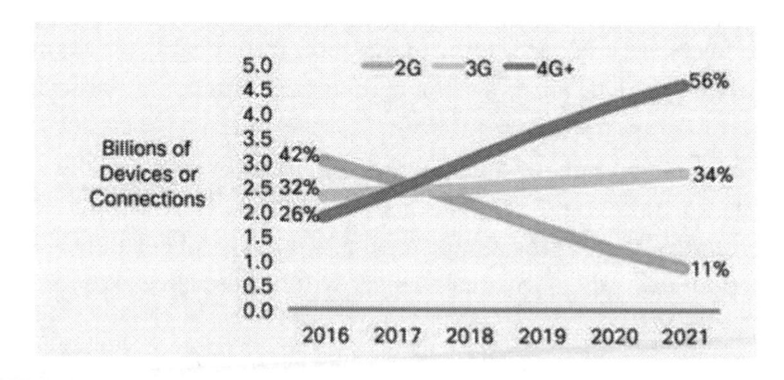

Figure 4.3 Global mobile devices (excluding M2M) by 2G, 3G, and 4G. (Source: [46].)

across the different generations, including, consequently, also the future evolution of 5G systems.

The following subsections analyze all the drivers for this network evolution, from a network technology point of view, starting from KPIs to the evolution of new user devices and terminals, and also describe the consequent impact on spectrum demand (regulated at both regional and global scales), which is on its turn determining the level of success of new communication technologies.

Needless to say that the evolution of these access technologies (especially from 4G toward 5G systems) is essential to enable the adoption of edge computing.

4.1.1 Network Performance Drivers for 5G Systems

From a general perspective, the key characteristics driving the current evolution of communication networks [108] are the following:

- 1,000 times higher mobile data volume per geographical area.
- 10–100 times more connected devices.
- 10–100 times higher typical user data rate.
- 10 times lower energy consumption.
- End-to-End latency of <1 ms.
- Ubiquitous 5G access, including in low-density areas.

These high-level requirements give the idea of the 5G performance expected at an operational level, referring to previous systems (4G). The reader may find this requirement very challenging. As can be easily seen in the following table (providing a set of KPIs for 1G/4G system), as a matter of fact, at every generation (more or less introduced every decade), the performance hop in terms of, for example, required peak data rate was exponential (also driven by the need to support new applications and services).

The KPIs required for 5G systems thus are a consequence of evolving market needs and are driven by the necessity to support even more challenging traffic flows, a bigger number of terminals, and even more diverse devices (this aspect will be discussed more in the next section).

More in detail, in 3GPP, a study has been done at the SA level about 5G service requirements. The objective was to develop

PARAMETERS	1G	2G	3G	4G
Introduced in year	1980s	1993	2001	2009
Technology	AMPS, NMT, TACS	IS-95, GSM	IMT2000, WCDMA	LTE, WiMAX
Multiple access	FDMA	TDMA, CDMA	CDMA	CDMA
Speed (data rates)	2.4–14.4 Kbps	14.4 Kbps	3.1 Mbps	100 Mbps
Bandwidth	Analog	25 MHz	25 MHz	100 MHz
Applications	Voice calls	Voice calls, SMS, browsing (partial)	Video conferencing, mobile TV, GPS	High-speed applications, mobile TV, wearable devices

high-level use cases and identify the related high-level potential requirements to enable 3GPP network operators to support the needs of new services and markets. Also, ITU defined in a recommendation [50] eight key "capabilities for IMT-2020," which formed a basis for the technical performance requirements [51] to be considered for communication technologies that aim at being able to fulfill the objectives of IMT-2020 systems (detailed requirements are described in Annex 4.A).

From Figure 4.4, the reader can understand very well that all the requirements for future communication systems will be very heterogeneous. As a consequence, it is obviously not possible to fulfill these challenging requirements at the same time, for all traffic streams. On the contrary, each service (e.g., eMBB, mMTC, URLLC) should be associated with a certain 5G network slice, capable of managing a certain traffic flow and its own specific subset of required KPIs (all listed in Annex 4.A for the sake of clarity).[1]

4.1.2 The Importance of New Devices for 5G Systems

We have seen how complex the evolution of the communication infrastructure is, and how the actual usage of 5G systems will be influenced and determined by the introduction of new devices in the market. Thus, we can say that the new 5G infrastructure is pushing into the market the urgent need for 5G-enabled terminals. On the other hand, we should also note that the introduction of new products and models of

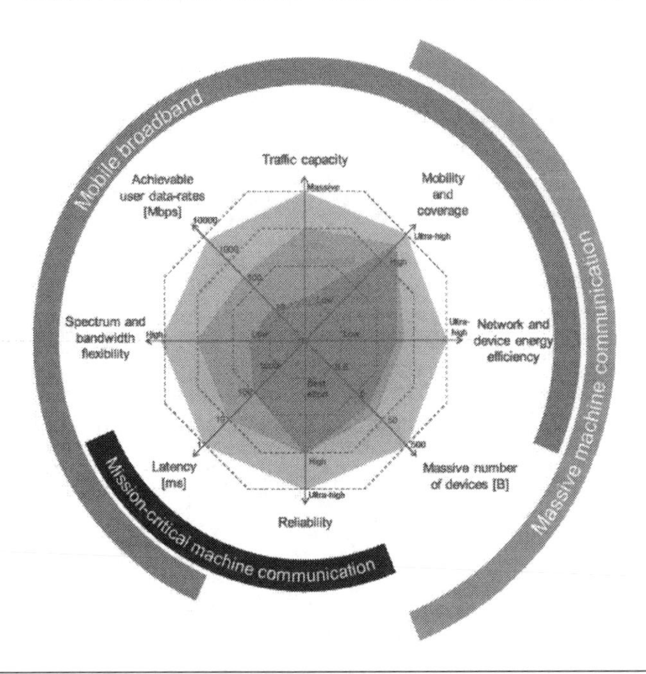

Figure 4.4 3GPP SA requirements spider diagram [52].

devices in the market often acts as a catalyzer for new services. This phenomenon is particularly true, if we think about the beginning of the smartphone era, characterized by the introduction of a new generation of devices, more capable and with bigger displays, but above all more usable, trendy, and appealing for the end user (and often also very expensive, compared to the previous generation of phones). The huge sales volumes (matching the interest of a consolidated customer base) generated a high number of connections and stimulated the actual usage of 4G systems. Furthermore, we can also say that the advent of smartphones stimulated the proliferation of an ecosystem of applications, fed by a growing community of developers, introducing a multitude of new services for end users. (Note: here, a careful reader may notice that 4G networks were not specifically designed for these services, which arrived later, and were instead driven by the advent of smartphones.)

Thus, we can essentially say that the beginning of the 4G success (in terms of global revenues, justifying all 4G investments after the 3G deployment) was also determined by the introduction of smartphones. Currently, for 5G systems, many experts are agreeing to consider vertical market segments (like automotive, industry 4.0, smart cities, ...)

as a key driver for the introduction of new services. But we should not forget here again the key role of new devices as a catalyzer. This time, heterogeneous terminals, but also wearable devices, sensors, and any sort of connected devices will be key for the actual success of 5G systems. Actual revenue streams and the related specific 5G business cases are still not clearly defined, at this point in time, but the importance of the role of devices appears to be clear in this case. In fact, if the actual 5G services are driven by vertical market segments (discussed more in Section 4.4.2), new-generation devices will need to support heterogeneous requirements, for example, those not only coming from traditional broadband connections between humans, but also related to emerging machine-to-machine communications, cloud robotics, factory automation, automated and connected driving, etc.

In summary, the typical innovation cycle is expected to be driven once again by new 5G devices, which can not only justify the usage of the new infrastructure, but potentially also drive the introduction of new services – even those services that are still not present today. In this sense, both 5G infrastructure and terminals should also be future proof and should flexibly accommodate the specific needs of new (and still unknown) services.

4.1.3 Spectrum Evolutions toward 5G Systems

Another key aspect for the success of communication systems is the possibility to have access to an increasing portion of the spectrum. This need is very critical for the introduction of every new generation (4G, 3G, 2G), and its relevance is evident especially for 5G systems, where the requested bandwidth is increased with respect to previous systems. Moreover, adding a new system on top of the existing range of technologies is always a critical aspect, as spectrum fragmentation (even at different regional levels) is a barrier for the global adoption of 5G, especially when customers need to connect while travelling and moving across different countries. Spectrum need is thus an asset that is carefully regulated by international institutions, taking into account many needs.

In particular, the following table summarizes the main GSMA's key 5G spectrum positions that focus on the areas where governments, regulators, and the mobile industry must cooperate to make 5G a success [53].

1 5G needs a significant amount of new harmonized mobile spectrum. Regulators should aim to make available 80–100 MHz of contiguous spectrum per operator in prime 5G mid-bands (e.g., 3.5 GHz) and around 1 GHz per operator in millimeter wave bands (i.e., above 24 GHz).

2 5G needs spectrum within three key frequency ranges to deliver widespread coverage and support all use cases. The three ranges are: Sub-1 GHz, 1–6 GHz, and above 6 GHz.
- Sub-1 GHz will support widespread coverage across urban, suburban, and rural areas and help support Internet of Things (IoT) services
- 1–6 GHz offers a good mixture of coverage and capacity benefits. This includes spectrum within the 3.3–3.8 GHz range, which is expected to form the basis of many initial 5G services
- Above 6 GHz is needed to meet the ultrahigh broadband speeds envisioned for 5G. Currently, the 26 GHz and/or 28 GHz bands have the most international support in this range.

3 The ITU World Radiocommunication Conference in 2019 (WRC-19) will be vital to realize the ultrahigh-speed vision for 5G, and government backing for the mobile industry is needed during the whole process.

4 Exclusively licensed spectrum should remain the core 5G spectrum management approach. Spectrum sharing and unlicensed bands can play a complementary role.

5 Setting spectrum aside for verticals in priority 5G bands could jeopardize the success of public 5G services and may waste spectrum. Sharing approaches like leasing are better options where verticals require access to spectrum.

6 Governments and regulators should avoid inflating 5G spectrum prices (e.g., through excessive reserve prices or annual fees) as they risk limiting network investment and driving up the cost of services.

7 Regulators must consult 5G stakeholders to ensure that spectrum awards and licensing approaches consider technical and commercial deployment plans.

8 Governments and regulators need to adopt national spectrum policy measures to encourage long-term heavy investments in 5G networks (e.g., long-term licenses, clear renewal process, spectrum roadmap).

From an edge computing perspective, the reader may argue that multi-access edge computing (MEC) success is not related to 5G spectrum regulations. Actually, this is not true. In fact, edge computing instances are by nature placed in close proximity to end users; as a consequence, in a world where communications are increasingly based on mobile networks, and where mobility is key, the success of 5G (and its availability at global scale) is also critical for the introduction of MEC. More in particular, the adoption of low-latency communications (i.e., new radio for URLLC in 3GPP) is especially important for delay-critical use cases driven by several vertical markets like industrial automation, cloud robotics, augmented and virtual reality, e-Health use cases, smart grids, autonomous driving, wideband PPDR, trunking, and smart manufacturing in the context of Industry 4.0. This new air interface often requires a licensed/dedicated spectrum and is able to ensure predetermined high levels of performance. Thus mobile operators are not the only ones now starting to look at the spectrum as an asset for introducing this network infrastructure (e.g., for connected and automated industries).

In addition, the usage of unlicensed spectrum and the huge and increasing presence of Wi-Fi networks and the related evolutions (e.g., Wi-Fi 5, Wi-Fi 6) are key for indoor/outdoor hotspot traffic scenarios like shopping malls, enterprises, or municipalities, or even trains and airplanes. Also, in these cases, the usage of spectrum for Wi-Fi connectivity is very important for the adoption of edge computing. New 5G systems will also have to better integrate cellular and WLAN connectivity at different levels (e.g., radio link aggregation, path aggregation) in order to maximize the usage of the spectrum to provide seamlessly better user experience and a low-latency environment to end customers.

4.2 The Need for the "Edge"

A key advantage for the edge computing adoption is given by its access-agnostic nature. In fact, MEC can be deployed with different communication technologies, spanning from cellular, but also Wi-Fi and fixed networks. In this perspective, MEC is key for 5G systems, which are expected to include over a unified core network the integration of LTE with New Radio (5G NR) accesses, also with Wi-Fi and fixed networks.

In the following section, we describe the main MEC drivers for 5G, from technical perspective and from a generic stakeholder perspective, highlighting some of the most important KPIs relevant for the introduction of MEC (Section 4.3 thus presents few selected use cases for 5G, while the specific technical status of 3GPP and other bodies are instead described in Section 4.4).

4.2.1 Key Drivers for MEC in 5G

The most commonly recognized key driver for the introduction of edge computing is the delay. Practically, all stakeholders agree that MEC is especially beneficial for the provision of low-latency environments, thanks to the introduction of cloud computing at the edge of the network, thus in close proximity to end users. This is true, but it is worth mentioning that delay is not the only KPI for MEC. Mobile operators and service providers are in fact progressively defining their strategy for the adoption of edge computing, and their positioning and motivation [54] for the adoption of edge computing is becoming more and more specifically defined and elaborated, with respect to past years.

Later in this book (in Chapter 10), we will analyze more in detail the operator profile, as a key stakeholder in the MEC ecosystem. Anyway, we can already anticipate that mobile operators and service providers are looking at MEC as a technology providing two kinds of benefits: *revenue generation* and *cost saving*. In particular, as infrastructure owners, they are interested in the following KPIs (described more in detail in Chapter 10):

- Delay improvement (enabling new services)
- Network utilization and cost savings (given by shorter data traffic paths)
- Energy efficiency and total cost of ownership (TCO)
- Management of computation and networking resources in network function virtualization (NFV) environments

In fact, most of the operators are conducting a long process of network transformation toward 5G, which includes a progressive shift to full virtualization. MEC is a key step in this path.

A careful reader would have noticed that this is of course a superset of KPIs derived from both *cost saving* and *revenue generation*. In fact, the assumption is that MEC deployment is mainly foreseen by stakeholders as coupled with NFV, that is, the introduction of edge computing and the related timing should be considered as strictly coupled with the progressive virtualization path of network infrastructure.[2]

4.3 Exemplary Use Cases for MEC

Use cases enabled by edge computing are many and heterogeneous, as they span across multiple market segments and related service opportunities for an entire and diverse ecosystem of stakeholders, ranging from fixed and mobile operators, service providers and OTT players, to application developers, cloud service providers, etc. In fact, ETSI ISG MEC have identified and described more than 35 use cases, conveniently classified into three main categories [58]:

1. Consumer-oriented services,
2. Operator and third-party services, and
3. Network performance and QoE improvements.

These categories are obviously only a convenient tool to classify some use cases identified so far and should not be seen as a limitation; in fact, a clever reader may have already noticed that some use cases could partially fall into more than one category, and also future services not listed and considered now could emerge later, for example, during a more mature introduction of edge computing in 5G systems and beyond. By the way, it is worth providing at least a rationale for these categories, though an exemplary use case, which could give an idea of the potential opportunities given by the adoption of MEC.

4.3.1 Consumer-Oriented Services

These are innovative services that generally benefit directly the end user, that is, the user using the UE. This can include gaming, remote desktop applications, augmented and virtual reality, cognitive assistance, etc.

In particular, the case of gaming (but also virtual/augmented reality) applications is particularly interesting in this early phase of 5G deployments, and also gives a concrete idea of services provided to the customer where E2E latency perceived by the end user is critical. In fact, gaming applications need short reaction times in order to provide the best performances. The software startup Edgegap (www.edgegap.com) proposes solutions to automate the decision-making process and deployment of gaming servers throughout hundreds of data centers. As a result, the best MEC server locations are determined, based on the end users' location, providing thus gaming applications running closer to their players (with obvious benefits with respect to online experiences given by public clouds). In addition, cost savings are realized by only running games when and where there is actual user demand.

4.3.2 Operator and Third-Party Services

These are innovative services that take advantage of computing and storage facilities close to the edge of the operator's network. These services do not usually necessarily benefit the end user, but in any case, can be usually operated by operators in conjunction with third-party service companies (e.g., active device location tracking, big data, security, safety, enterprise services).

The example of the active device location tracking use case is particularly meaningful due to the benefits derived from the usage of local information to provide added value service. In fact, a geolocation application hosted on the MEC server using real-time tracking algorithms for terminal equipment (based on GPS and network measurements) can provide an efficient and scalable solution with local measurement processing and event-based triggers, and enables location-based services for enterprises and consumers (e.g., on opt-in basis), for example, in venues, retail locations, and traditional coverage areas where GPS coverage is not always available. A classic example of application services includes mobile advertising and proximity marketing (www.vividaweb.com), where contents can be generated at the MEC and potentially also customized for the user, dependently on his profile, and also based on the specific location where he connects to the network.

4.3.3 Network Performance and QoE Improvements

This third category is mainly interesting for operators and infrastructure providers, since it includes services generally aimed at improving performance of the network, either via application-specific or generic improvements. The user experience is improved, often in a transparent way to the end user, but in practice, these are not new services provided to the customer (examples of typical use cases are content/DNS caching, performance optimization, and video optimization). Here, mobile operators and/or content distribution network (CDN)-based contents providers can offer better video streaming services to premium users, for example, with lower delay or packet error rate at application level, thus with better performance and without the need to create new services.

A relevant example for performance improvement is given by MEC testing performed by Vodafone and Saguna Networks.[3] In this lab setup, they tested the impact of MEC on improving video streaming quality of experience. The tests involved comparing video streaming from an edge virtual video server to a virtual video server using a remote Amazon Web Services (AWS) in an emulated mobile network environment (with data modeling the typical latency and congestion experienced by users on mobile networks in the UK). Performance assessment[4] was conducted in terms of the following three KPIs:

- Time to start: the time interval between clicking play and when the video starts playing.
- Number of stalls: a numerical count of how many times the video pauses for re-buffering.
- Waiting time: the total time spent waiting, which is calculated by adding all the re-buffering pauses.

4.4 Edge Computing: 5G Standards and Industry Groups

As already anticipated, we often use the acronym MEC in this book as a synonym of edge computing, for the sake of simplicity. On the other hand, the acronym is also referred to the work done by ETSI ISG MEC [10][15] to define a standard that is (as a matter of fact) the only international standard available for edge computing. From this perspective, we could simply continue using these two terms as synonyms. On the other hand, it is also important to clarify that 5G system architecture is defined by 3GPP, and the definition of edge computing in this perspective starts from a more general point of view. Then by nature 3GPP is still considering edge computing from a more generic point of view with respect to ETSI standard. Thus, it is worth clarifying that current standardization activities (both in MEC and 3GPP) are still ongoing, as the related support of edge computing in 5G system still needs to be fully specified and defined. The following sections provide an updated overview of the current status of activities.

4.4.1 3GPP Standardization Status

The introduction of edge computing in 5G systems has been initiated by 3GPP in the SA2 working group in 2017 (and targeting Rel-15 specifications). This was a pivotal step toward the integration of this key enabler in 5G. In fact, in the TS 23.501 specification [42], defining the System Architecture for the 5G System, edge computing is clearly indicated as a technology that "enables operator and 3rd party services to be hosted close to the UE's access point of attachment, so as to achieve an efficient service delivery through the reduced end-to-end latency and load on the transport network".

One clarification is required here: a careful reader of this 3GPP specification could notice that the introduction of edge computing is quite generic. This is perfectly aligned with the fact that the TS 23.501 actually defines 5G standard at "Stage 2" level, which means that more detailed information on implementation aspects (thus at "Stage 3" level, according to the 3GPP terminology) is expected in other specifications, while the SA2 group only defines higher level requirements and technical enablers (thus, not detailed implementation). For that reason, more work is expected to better specify the support for edge computing in 5G. Interested readers could consider as reference the current work of other groups in 3GPP, which have recently started working on edge computing (always from a 3GPP perspective):

- SA5: "Study on management aspects of edge computing" [63]; this report is focused on 3GPP management system and the related support for the deployment and management of edge computing;
- SA6: a more recent "Study on application architecture for enabling EDGE applications" [64]; this report aims at evaluating the need for an overall application framework/enabling layer platform architecture and associated requirements to support Edge computing in 3GPP networks, and identifies key issues and potential solutions to support deployment of edge applications (e.g., through discovery and authentication), including potential UE and network edge APIs.

In addition to that, SA2 started a new study item (called "Study on Enhancement of Support for Edge Computing in 5GC" [62]), targeting two main objectives:

- to study the potential system enhancements for enhanced edge computing support (e.g., discovery of IP address of application server, improvements to 5GC support for seamless change of application server serving the UE)

- to provide deployment guidelines for typical edge computing use cases, (e.g., URLLC, V2X, AR/VR/XR, UAS, 5GSAT, CDN)

This last study (which will be published in a TR 23.7xy and will target Rel.17) is very important for the ecosystem because it will help stakeholders (and especially infrastructure owners) to clarify constraints and trade-offs related to edge computing deployment options. Moreover, the study already clarifies that some additional aspects related to the application layer architecture for enabling edge computing are in the scope of SA6; thus we can expect some 3GPP work in this area too.

In any case, the 3GPP approach is always meant to be generic, as this SDO tends to consider in principle many use cases, and many possible implementations of edge computing. For that reason, ETSI MEC standard (shortly described in Section 4.4.3) is only one option; for example, in addition to proprietary solutions, open source implementations, or even specific solutions driven by industry groups. Of course, the general value of standards for interoperability is undoubtable (and here ETSI MEC plays a key role, as the only international standard available in the space). But the contribution of industry groups and open source communities in this space is essential to drive innovation and adoption of this technology. All in all, MEC is cloud computing at the edge, so the engagement of software developers is essential (and traditionally, these communities can be found outside SDOs too).

4.4.2 Industry Groups

Edge computing facilitates enhancements to the existing applications and offers tremendous potential for developing a wide range of new and innovative 5G services (e.g., automotive, industrial automation, multimedia, e-Health, smart cities, virtual/augmented reality) by enabling authorized third parties to make use of local services and computing capabilities at the edge of the network. In this sense, edge computing enables many vertical market segments, which in principle are very heterogeneous, driven by different business needs and imposing different requirements to the communication network. To better drive these specific technical requirements, specific skills and competencies are thus required from the different verticals, and this kind of expertise is generally not found in 3GPP groups (traditionally focused on communication network, from radio to core) nor in ETSI ISG MEC (which defines the general aspects of cloud computing infrastructure for the edge). Thus, these standard organizations need to be complemented by specific industry groups, driven by key stakeholders in the respective verticals. For example, we can mention:

- 5GAA (5G Automotive Association), www.5gaa.org
- AECC (Automotive Edge Computing Consortium), https://aecc.org/
- 5G-ACIA (5G Alliance for Connected Industries and Automation), www.5g-acia.org/
- VR-IF (Virtual Reality Industry Forum), www.vr-if.org/

As one example for all, 5GAA is a huge industry association putting together automotive stakeholders and communication network experts. These two big communities are working together to develop end-to-end solutions for future mobility and transportation services. In order to do that, of course the automotive industry needs the expertise and technical skills coming from ICT stakeholders, and vice versa (as communication network experts are not familiar with the specific problems coming from automotive players and specific in-car implementations). The 5GAA association is thus driving specific use cases of interest, defining technical solutions to implement them in 5G networks, and building a relevant consensus to push the identified solutions to relevant standard bodies (e.g., 3GPP for the radio and core network aspects, of ETSI MEC for edge cloud standardization).

More in detail, since edge computing is one of the six priority areas for 5GAA, this association published a white paper [65] on edge computing for advanced automotive communications. This work provided an overview of selected automotive use cases (as introduced by 5GAA), showing how edge computing (and in particular, also standardized solutions) can be considered as a key technology supporting multiple services for connected vehicles, especially in multi-operator, multi-vendor, and multi-OEM systems.

Overall, the role of verticals and related associations (pushing their specific needs and technical skills) is thus key for the success of edge computing, as well as standard groups.

4.4.3 The Role of ETSI MEC in 5G

We have clarified several times that ETSI MEC standard is access-agnostic, thus in principle independent from a specific deployment over a specific communication network (e.g., 4G, 5G, or even Wi-Fi and fixed network). Moreover, from a principle perspective, the MEC standard should support many options for "edge" implementation (in the sense that MEC architecture of an MEC server is basically the same, while "edge" locations of MEC servers should be all possible, in principle, according to the standard). Nevertheless, when it comes to MEC in 4G and 5G networks, specific clarifications on the actual deployment options are needed (e.g., where is the edge, how to insert an MEC server within the cellular network, how the traffic redirection and radio/application mobility is managed). For this purpose, a couple of ETSI white papers have been published, to shed some light on these aspects:

- "MEC Deployments in 4G and Evolution towards 5G" [66]: this paper explored how the MEC system can be deployed in existing 4G networks, by showing different options to install the MEC host along with the 4G system architecture components (e.g., the so-called "Bump in the wire" option, or other deployments over distributed EPC; e.g., distributed SGW with local breakout (SGW-LBO), or again with control/user plane separation (CUPS)). This study also observed the impact of such installation choices on the running system and architecture, and showed a possible path for migration to future 5G networks, looking at the problem from different angles, including compliance with 5G system architecture, adoption of cloud computing and NFV paradigm, and protection of the investment during network upgrade.
- "MEC in 5G Networks" [67]: this paper constituted a follow-up of the previous study. It provided some example scenarios of physical deployment ((i) MEC and the local user plane

functions (UPF) collocated with the base station; (ii) MEC colocated with a transmission node, possibly with a local UPF; (iii) MEC and the local UPF collocated with a network aggregation point; (iv) MEC collocated with the core network functions, i.e., in the same data center). Then, moving from physical to logical implementation, the paper illustrated the key components involved for an integrated MEC deployment in 5G networks, for example, by discussing the ability of MEC, seen as a particular implementation of a 5G AF (Application Function) in 3GPP, to interact with the 5G system to influence the routing of the edge applications' traffic and the ability to receive notifications of relevant events, such as mobility events, in the 5G system for improved MEC deployment efficiency and end user experience. Moreover, the paper also presented the benefit of deploying MEC in Local Data Network, made possible by the flexibility in locating 3GPP UPFs, which are implementing the data plane for MEC hosts.

Starting from the latter white paper, due to the huge interest for MEC in 5G deployments, recently, ETSI ISG MEC also started a work item [68], with the intention to document the key issues and solutions for MEC integration in 5G networks. In particular, this document discusses issues related to MEC Application Function C-plane interactions with 5GC, including the mapping of MEC procedures to procedures used in 3GPP 5G system, options for functional split between MEC and 5G Common API frameworks, organization of MEC as Application Function(s) of 5G system, and pertinent interactions with the (R)AN. In addition, this work item aims to identify any yet missing 5G system functionality, for example, to provide indications about future standardization work needed. This is at the end a key work for ETSI MEC standard, because in principle, some support for 5G (dually to the study in SA2) could be foreseen in the ISG.

4.5 MEC and Network Slicing

Network slicing is a key concept for 5G networks, initially introduced by NGMN [69] and then in 3GPP specifications[5] for Rel.15 and beyond. It is essentially defined as a partitioning of the physical network into multiple virtual networks customized to meet diverse set of vertical requirements. A network slice is thus an actual portion of 5G network, including all network elements (from RAN to core), and it is instantiated by the system management with specific characteristics coming from the particular service/traffic to be served. An example of network slicing, related to automotive field, is the following:

- The vehicle may need to simultaneously connect to multiple slice instances, belonging to different slice/service types (SSTs), to support different performance requirements of multiple automotive use cases.
- For example, Software Update and Tele-Operated Driving use cases could be associated with eMBB slice and URLLC slice, respectively, based on their KPI requirements.

But, how to slice this "5G cake"? And, what is the role of MEC in network slicing? In order to answer this relevant question, we should first have a better look at the KPIs, for example, the E2E delay. In this case, 3GPP doesn't take care of the real E2E performance requirements,[6] and thus MEC should be considered, as in principle it is really outside the domain of 3GPP responsibility, in that perspective. In other words, while the network slicing mechanism is managed within 3GPP and involves 3GPP entities, MEC apps are the actual user traffic end point, and the MEC system should be considered as well when talking about real E2E performances.

For this reason, the MEC study on network slicing [74] discusses how to "slice the 5G cake" and the related impacts on MEC, by taking into account E2E service requirements on the quality of service (QoS). The work focuses on identifying the necessary support provided by MEC for network slicing and, in addition, determining how the orchestration of resources and services from multiple administrative domains could facilitate that.

As we have anticipated, network slicing is a key concept for 5G. In principle, it widely covers many verticals, and is also related to service level agreements (SLA) between different stakeholders (e.g., between road operators/car makers and mobile operators/service providers). For this reason, 3GPP does not cover all implementation aspects related to network slice requirements and parameterizations.[7]

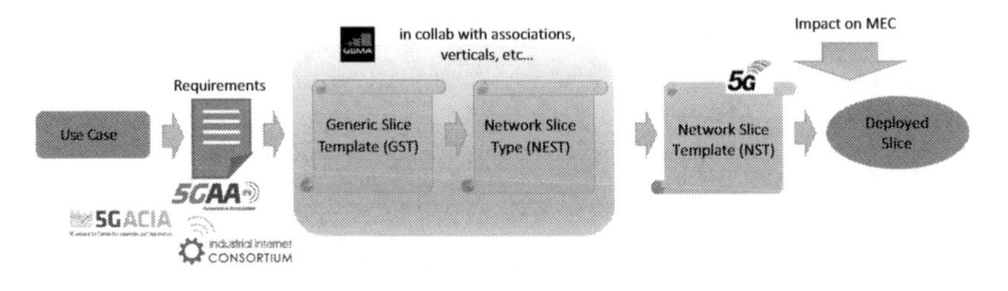

Figure 4.5 Example of process flow for the definition of a network slice, and possible impact on MEC. (Elaboration from GSMA.)

Moreover, high-level characteristics and parameterizations related to network slices are introduced in GSMA network slicing task force (NEST)[8] [75], which starts from a collaboration with associations and verticals, to gather requirements, and produce relevant generic slice templates (GST), containing attributes (e.g., throughput, max UP latency) and related performance (e.g., isolation, security model, mobility support) expected at E2E level, for the instantiation of a network slice. Starting from these high-level configurations, a network slice template (NST) thus identifies the 5G slice in the 3GPP domain (see Figure 4.5).

As discussed before, in order to meet E2E performance requirement, this slice instantiation should be also properly coupled with MEC App instantiation. This work potentially influences 3GPP and ETSI MEC, for example, starting from the current studies (in the two respective bodies) related to MEC in 5G.

Annex 4.A – IMT2020 Systems: Minimum Technical Performance Requirements [51]

TECHNICAL REQUIREMENT	USAGE SCENARIO APPLICABILITY				TARGET VALUE
	EMBB	MMTC	URLLC	GENERAL	
Peak data rate	✓				DL: 20 Gbps UL: 10 Gbps
Peak spectral efficiency	✓				DL: 30 bps/Hz UL: 15 bps/Hz
User experienced data rate	✓				DL: 100 Mbps UL: 50 Mbps

(*Continued*)

TECHNICAL REQUIREMENT	USAGE SCENARIO APPLICABILITY				TARGET VALUE		
	EMBB	MMTC	URLLC	GENERAL			
5th Percentile user spectral efficiency	✓				**TE**	**DL (bit/s/Hz)**	**DL (bit/s/Hz)**
					InH	0.3	0.21
					DU	0.225	0.15
					RU	0.12	0.045
Average spectral efficiency	✓				**TE**	**DL (bit/s/Hz)**	**DL (bit/s/Hz)**
					InH	9	6.75
					DU	7.8	5.4
					RU	3.3	1.6
Area traffic capacity	✓				10 Mbit/s/m^2		
User plane latency	✓		✓		URLLC: 1 ms eMBB: 4 ms		
Control plane latency	✓		✓		20 ms (10 ms encouraged)		
Connection density		✓			1 million devices / km^2		
Energy efficiency	✓				Qualitative measure Support a high sleep ratio and long sleep duration		
Reliability			✓		1×10^{-5} success probability for TX 32B in 1 ms		
Mobility	✓				**TE**	**Mobility (km/h)**	**TCDL (bit/s/Hz)**
					InH	10	1.5
					DU	30	1.12
					RU	120	0.8
					RU	500	0.45
Mobility interruption time	✓		✓		At least 100 MHz, up to 1 Gbps for higher freq. bands		
Bandwidth				✓			

Notes

1 Note: the concept of network slice is not deeply discussed in this book, but the interested reader can find relevant resources in 3GPP specifications for 5G system architecture [70].

2 In principle, the MEC architecture [59] also permits stand-alone implementations, of course based on virtualized infrastructure, but not necessarily in NVF environments [60]. The second possibility is better clarified and defined in the second phase of MEC standardization [61].

3 See Vodafone's blog post: www.vodafone.com/content/index/what/technology-blog/video-streaming-experience.html#

4 See also Saguna website: www.saguna.net/blog/improving-video-streaming-customer-experience-with-multi-access-edge-computing-research-results-from-vodafone-and-saguna/

5 Interested readers can refer to the following 3GPP specifications, relevant to network slicing: TS 23.501 (on 5G System Architecture, ref. [70]), TS 22.261 (on 5G Requirements, Ref. [71]), and TS 28.531/28.532 (on 5G Slice Management, Ref. [72,73]).

6 In fact, according to 3GPP specifications, the packet delay budget (PDB) defines an upper bound for the time that a packet may be delayed between the UE and the UPF. This means that the N6 reference point toward the DN (thus until the MEC App) is not included, by definition, is this delay budget. On the contrary, E2E performances requirements should be determined by considering the overall path of user traffic packets, thus including N6 and the MEC app.

7 In TS 28.531, an NST is defined by 3GPP as a subset of attributes' values used for creation of instances of network slice information object class (IOC). The content of NST is not planned to be standardized by 3GPP, that is, it is defined by MNO and vendor.

8 www.gsma.com/futurenetworks/wp-content/uploads/2018/07/1_2_GSMA-Progress-of-5G-Network-Slicing_GSMA-NEST_vice-chair.pdf

PART 2

MEC AND THE MARKET SCENARIOS

5

The MEC Market

The Operator's Perspective

This chapter discusses the benefits of multi-access edge computing (MEC) from a telecom operator perspective, including business aspects, deployment considerations (MEC, software-defined networking [SDN], network function virtualization [NFV], ...), edge disaggregation, and distributed operating systems.

5.1 What Does MEC Mean to Operators?

5G promises advantages and capabilities operators know they will need to compete globally, open new lines of revenue, and expand into new industries. But while early adopters race toward imminent, albeit limited, deployments, most operators are still plotting a path to 5G. This is largely an unchartered territory. Supporting technologies and ecosystems are only just being formed. Yes, consumer and industry demands are fast outgrowing and will soon outpace the LTE networks by which they are currently supported. They are eager for 5G.

MEC represents the solution to these challenges, creating an opportunity right now to support countless new industry-driven and operator services, expanding the revenue potential of mobile networks and opening new avenues of business operators not previously explored.

MEC does not require operators to recreate the wheel. It is not a new box or system. Instead, it's a platform to aggregate a range of edge-based capabilities that are ready for deployment. Like a child discovering a drawer full of LEGO, you don't always know what you have until you start experimenting. As operators begin to experiment with MEC, they discover practical applications and new services that can power growth.

MEC bridges the gap to 5G without slowing progress along the journey. In this chapter, we delve into MEC's true value to operators, including its ability to supercharge today's 4G networks. We will explore its ability as a platform to deliver differentiated services based on low latency and enhanced quality of service. We will also discuss its value as a stepping stone to 5G-driven architecture, while helping to develop software infrastructure skills and expertise within operator teams.

5.2 Benefits of MEC

There has long been a desire in telecom to relocate computing functions ever closer to users. In fixed networks, operators have built content distribution networks (CDNs) or worked with CDN partners to deploy distributed network capabilities, caching popular content closer to where they are consumed. CDNs offered a way to improve customer experiences via faster content access and cost economics through de-duplicated traffic delivered over backhaul transmission lines.

Mobile operators have made attempts to integrate CDNs into their networks. Before MEC, some operators deployed servers directly on cell sites, essentially deploying compute and storage functions right alongside base stations. While the hope was to deliver more tailored content to certain locations, these efforts were not met with broad success. Operators discovered that "location-specific" services could be easily delivered from existing centralized locations, provided the end points could offer location information, such as in the case of smartphones equipped with GPS.

5.3 Igniting an Industry

Although early experimentation with pushing services closer to users brought mixed results, mass interest in finding ways to explore location-specific services that benefit from closer proximity to end users persists. Once implemented, services that range from traditional over-the-top (OTT) to augmented reality, mobile gaming, and connected vehicles have the potential to emerge as key differentiators for operators.

For example, the Internet of Things (IoT) is already a reality for many industries. It demands increasingly lower latency and predictable mobile services to support changing requirements. These requirements cannot be satisfied by the traditional data services widely available today. Analysys Mason forecasts that the total addressable mobile operator revenue from the IoT value chain will be $201 billion by 2025, and the MEC provides an opportunity to start capturing a share of this growth now.

Following years of experiments, the industry at large is coalescing around approaches to edge computing, following key developments:

- **Evolution of virtualization technologies** that has seen traditional network functions deployed as software. This results in edge computing resources becoming not just overhead infrastructure that sits on top of proprietary network equipment, but core infrastructure that also serves existing network functions.
- The type of **content consumed on mobile networks has evolved** from pure browsing and file downloads to more demanding and bandwidth-heavy applications, like HD video streaming and online collaborative gaming. The ability to properly manage new application types at the highest quality and consistency requires edge intelligence and localized services. With the assistance provided by real-time access network resource availability information, especially mobile radio resources, operators will be able to provide the right level of service to a heterogeneous landscape of applications. High demand makes it easier to develop a business case.
- New solutions have given operators the **ability to understand the available access capacity and demand in real-time** and allocate network resources in the most efficient way to deliver better quality of experience (QoE) and reduce infrastructure costs. Unlike the traditional, core network traffic management approaches that apply policies across

the whole network without the knowledge of edge resource capacity, MEC platforms with radio network information services (RNIS) can manage various applications according to real-time radio resources, taking action only during congestion periods. Some MEC platforms are capable of breaking specific application traffic out of its normal path and directing it instead to edge-based processors to meet service demands.

- **Internet of Things (IoT) adoption** is introducing billions of new devices to the network, with Gartner predicting as many as 20.4 billion by 2020.[1] In machine communication, these devices typically communicate with each other at close proximity and are best facilitated by the edge. This requires infrastructure that can identify and process traffic with optimal latency. Although solutions are already available for certain industries in closed environments like corporate and public locations, the greater opportunity is to extend this capability across all access technologies, including mobile, where latency is significantly higher and extremely variable, posting challenges for mission-critical machine communication. For instance, in autonomous driving scenarios, cars must communicate critical information within milliseconds so that other cars can react in real-time. Constant and consistent service is vital in environments like this.
- **Video image processing for applications** like security is gaining momentum in many industry verticals due to the ability to quickly identify specific behavior or activity that can be automatically followed by real-time actions, with key information converted to metadata. AR/VR is another example, following early implementations based mostly on the power available in mobile handsets. Next-gen services will require substantial and expensive processing power that is high in energy consumption, but building this into devices themselves implies bulky form factors that are nonstarters for some applications. Deploying processing in centralized cloud environments introduces too much latency and requires significant backhaul. All eyes are on the edge to clear these hurdles.

These developments can't happen seamlessly and in tandem without the guidance of defined industry standards. According to ETSI, standards provide:

- **Safety, reliability, and environmental care.** Telecom operators perceive standardized products and services as more dependable, which raises user confidence, and increases sales and adoption of new technologies.
- **Support of government policies and legislation.** Standards are frequently referenced by regulators and legislators to protect user and business interests, and support government policies. For instance, standards play a central role in the European Union's policy for a single market.
- **Interoperability.** Having a thriving ecosystem of devices that work together relies on products and services that comply with standards.
- **Business benefits.** Standardization provides a solid foundation for development of new technologies, opening market access, providing economies of scale, fostering innovation, and increasing awareness of technical developments and initiatives.
- **Consumer choice.** Standards provide the foundation for new features and options, contributing to the enhancement of consumer lives. Mass production based on standards provides a greater variety of accessible products to consumers.

5.4 Enabling Greater Value

A good example of the power of standardization is ETSI's Global System for Mobile communications (GSM) standard, first deployed in 1991. 3G and 4G would follow, underscoring the key role ETSI plays as a standards body to help drive the continued evolution of telecommunication networks, like NFV, SDN, and MEC.

MEC, NFV, and SDN will be important contributors to the success of future telecommunication services. According to McKinsey & Company,[2] the past several years have challenged telecom companies, with a 6% annual decrease in revenue and cash flow since 2010.

Global industry cash flows (EBITDA – capital expenditures),[1] $ billion

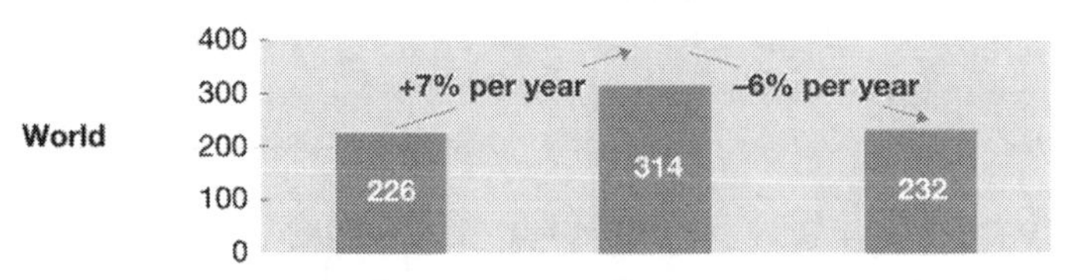

Capital-expenditure/revenue ratios for top telecom companies,[2] %

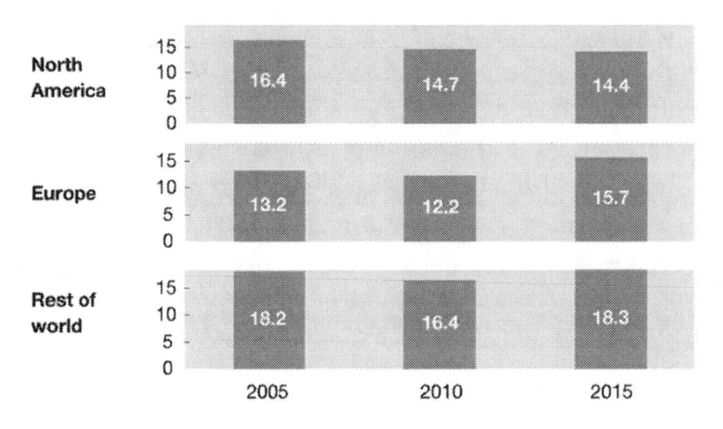

[1]Largest 250 telecommunications companies; EBITDA = earnings before interest, taxes, depreciation, and amortization.
[2]Largest 6–7 companies in each region.

McKinsey&Company | Source: S&P Capital IQ; McKinsey analysis

Telecom companies can alleviate the squeeze on margins and create more value by adopting next-generation technologies like MEC, NFV, and SDN. The adoption of these digital technologies and processes can also streamline business functions and improve customer satisfaction.

MEC platforms allow applications to be deployed at the edge, receive RNIS-relevant network information, and directly request network services. This is a departure from the standard architectures that have remained mostly unchanged going as far back as the introduction of GSM and continuing through to 3G and 4G networks. As a result, application demand can be addressed more intelligently, taking into consideration real-time radio conditions and allowing the applications to interact with the network infrastructure to request necessary services.

The ability to granularly address traffic demand on a mobile radio cell is a major shift for operators. Until recently, mobile networks have been primarily architected to support voice services with prioritized QoE and data services supported by centralized functions for distinguishing and managing traffic based on application types. The growing problem with this approach, in an age of diverse data demand, is that it treats the radio access network (RAN) as a generic pipe, providing average QoE. In reality, what happens is that some cells are overutilized and underutilized based on the time of day and location. It may seem intuitive to try managing this pipe as a hole to reduce the load on overutilized cells, but this approach also reduces the load on all other cells too, resulting in wasted resources.

By adopting MEC, operators can start taking varied approaches to traffic management in 4G networks while creating a platform that fosters experimentation with new services that mimic 5G capabilities, such as low-latency services and network slicing.

MEC has caught the attention of the broader operator community, and many operators around the world have active lab and field trials. Still, there are a number of constraints limiting the mass rollout of MEC:

- **No "killer" use case**. Although every generation of 3GPP mobile architecture has been accepted as part of the Telco infrastructure evolution, developments outside this body have required solid business cases. MEC is no exception, and before adopting it, operators want to understand how they will recover their technology investment.
- **MEC deployments require compute resources at deployment locations**. MEC platform suppliers could technically provide and manage the physical infrastructure but the greater opportunity is tied to full integration with the operator's NFV infrastructure. The process of building and decentralizing network data centers has begun among top operators, but the process is moving slowly and is limited to a dozen or so core locations that cannot fully exploit MEC's capabilities, such as delivery of low-latency applications.
- **We're not in a centralized world anymore**. All existing planning and operational procedures are built on the premise that all data traffic is transported to central points where it will be processed and routed to the Internet. In a decentralized

world, these processes need to be reviewed given traffic will be managed in different edge locations and based on usage that varies over time and per location.

- **Multi-vendor approaches** are a departure from the norm. Although MEC is being developed to provide a platform that enables different applications from potentially different vendors, operators are accustomed to sourcing vertical solutions from the same vendor. Building a multi-vendor platform and developing solutions that rely on best-of-breed applications require Telcos to take a new approach in their work with suppliers.
- **MEC must support true mobility**. Support for mobility is challenging for edge-based applications. For instance, what is the impact to a certain application when a user moves from an MEC domain to another whilst being served locally? This requires applications to be aware of mobility and adjust accordingly when a user is about to change MEC domains. Also, how can application developers begin to address the challenge? A first step to application developers gaining a real understanding of MEC, it's capabilities and challenges will come only when operators begin to deploy MEC across their infrastructure.

5.5 Business Benefits

The move to software infrastructure is necessary to support 5G promises via edge computing and software functions. Telcos are already familiar with the cost benefits of adopting a software infrastructure based on NFV with decreased CAPEX and OPEX. Most of the previous implementations have been on the core of the network and vCPE enterprise access infrastructures. There is still a great potential to reduce the cost by further automating infrastructure management and adopting elastic services that scale the physical resources up or down and make them available for other applications. However, these are just the low-hanging benefits of splitting hardware and software.

With the extension of computing resources toward the edge, operators can expand these benefits to provide low-latency edge services in locations where they are needed. We are no longer talking just about cost savings. This will also unlock tremendous potential for new applications and revenue streams.

5G will require a major change in how Telcos plan and operate their networks. In the meantime, operators don't have to wait. If they adopt MEC today, they can benefit immediately and get 5G-like benefits on 4G networks. Why wait for 5G when existing networks have the potential to deliver more value, and operators can get more from what they currently have without massive investments?

Building a platform that keeps existing architectures intact but provides more capabilities is advantageous for accommodating more capabilities that may be added in the future. This applies not just to technology, but also to people, skills, and business processes. This is particularly notable in an era when adding physical radio capacity can take up to 18 months, despite many expecting capacities on demand.

Typically, one-third of operator network costs are spent on real estate to house infrastructure. This includes tens of thousands of physical locations that host radio equipment, transmission aggregation, and core services like voice and data processing. There is significant potential to transform most of these locations into edge compute infrastructure that can monetize idle processing and storage capabilities due to how close they are to the user versus a public cloud data center.

When evaluating the potential to monetize MEC, operators will likely find the following business models most appealing:

- **Dedicated edge hosting**. The operator manages edge-located compute and storage resources, making it available to partners like third-party application developers. In this scenario, the operator is only responsible for ensuring that traffic is managed and delivered, while the developer takes ownership of assuring all other aspects of the service. An example could be a CDN operator that wants to move services even closer to end users.
- **XaaS**. The Infrastructure-as-a-Service, Platform-as-a-Service, and Network-as-a-Service approaches see the Telco operating similarly to a cloud provider, delivering compute and storage capabilities, APIs, and virtual network functions through a cloud portal. The operator may provide a service that is akin to what is offered by public cloud providers, but with the

advantage of lower latency experiences and integration with a range of MEC capabilities. Consider a CDN leveraging real-time information about radio conditions to choose the best content quality to deliver.

- **Systems integration**. The operator builds upon an existing systems integration business and offers turnkey solutions with integrated MEC capabilities. An operator may use this strategy to deliver IoT services to enterprise clients that require low latency, and also complement MEC capabilities with additional offerings to meet the needs of other industry verticals, like automotive, smart cities, and health care.
- **Business to business**. The operator provides edge processing and interfaces to offer edge-enabled solutions to enterprise customers that build solutions on top of it. This model could apply to the previous systems integration examples only to the extent of providing common services and interfaces to all customers from a given industry.
- **Traditional retail**. The operator develops its own services, such as mobile gaming, video surveillance image processing, or augmented reality to augment basic voice and data services.

5.6 Finding the Network Edge

One of the fiercest debates raging in telecom is where exactly the edge of the network is located. The short answer is, it depends.

The operators really want to understand where the equipment will be deployed so that they can start gaining an understanding of how it will be managed. There is often a misconception that edge equipment must be installed at every cell site, but this isn't necessarily true. For instance, in the case of traffic management, MEC can be deployed deeper into the network at aggregation sites. On the other hand, where ultralow latency is the goal and the intent is to keep as much traffic off of the network as possible, deploying right at the base station may make sense. Of course, there are high costs associated with this approach given how many more implementations are required. Fortunately, most MEC applications do not require deployments at the extreme edge of the network.

In many cases, it would be more appropriate to deploy at either:

- **Central offices** that are distributed across hundreds of locations for a typical large operator and situated halfway between the access and core networks, cutting latency of traditional applications in half. These locations often host transmission aggregation equipment with a footprint large enough to accommodate small data centers; or
- **CloudRANs** that are becoming more commonplace following the disaggregation of the RAN and operator initiatives to centralize baseband units (BBU) to virtualize and pool resources from remote radio heads (RRH) that remain at individual cell sites. The result for a typical operator is thousands of new BBU locations being built within a country. Because these locations are designed from scratch and based on generic computing standards, they make a perfect location to equip with MEC platforms and services.

Typically, it will be less complex and more cost-effective to deploy MEC at new, software-based locations that are being freshly architected. It is more challenging to deploy on existing physical infrastructure built several decades ago to host specific network switching and routing equipment. These locations were typically architected to address more limited, DC-based power consumption needs. The move to a data center model that hosts compute and storage equipment brings increased density of equipment per square meter and significantly increased, AC-based power consumption needs, generating more heat. For all of these reasons, deploying on traditional physical infrastructure, while not impossible, is certainly challenging from a cost and complexity standpoint.

Fortunately for operators, virtualization movements are well underway, with NFV commonplace and virtual infrastructures built into most technology strategies. While virtualization doesn't bring all of the baggage of building on top of physical networks, deploying a decentralized virtual infrastructure requires careful planning.

NFV standards continue to evolve. In this process, they provide the necessary components to automate the deployment and scaling of network functions to increase the efficiency of virtual infrastructure usage. Still, most NFV deployments have been implemented on central locations with data centers large enough to scale the physical compute resources based on capacity needs. The same flexibility is not available when moving applications to edge locations. As explained earlier in this chapter, these are locations closer to user access, in the magnitude of hundreds or thousands per country. They are also much smaller in terms of footprint, making the maximum compute capacity and its efficient use of the utmost importance.

Due to their early initial availability, hypervisor-based virtual machine solutions have been the default choice for NFV deployments. But these machines bring substantial data footprints and take minutes to boot, posing challenges in environments with limited resources, resulting in an emerging preference for container-based virtualization solutions. While these solutions can better scale for more efficient computing, they also require more advanced orchestration.

Many Telcos have started to demand cloud native applications for better efficiency and compatibility with new virtual infrastructure. An example of this are the container-based NFVs that leverage hyperscaling capabilities, recently developed by cloud providers like Google and Amazon. Containers bring much less overhead than virtual machines and can be deployed faster. However, they also carry new security considerations and network challenges that virtual machines do not. Containers pose a security risk because they run directly on server hardware. For instance, a hacker could take advantage of a compromised container to break gain access to an entire network.

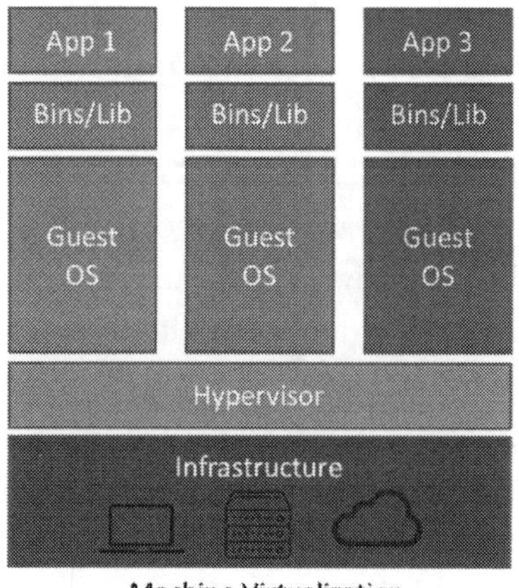

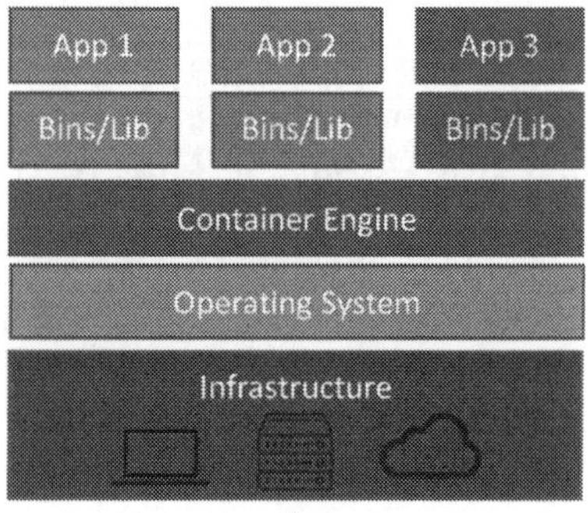

The operators must take into consideration that the infrastructure they evaluate takes time and can slow the advancement of edge cloud strategies. This is the reality of software-based infrastructures, where the pace of evolution is faster and more conducive to incremental steps and hybrid approaches. Different network applications will bring different requirements, whether related to packet processing performance, security, or traffic segregation. While network functions that process user data might benefit from hypervisor-based robustness, other functions that process network signaling and control traffic might work better on a hyperscaling environment like containers.

MEC platform deployments require NFV and an underlying compute and storage infrastructure. Although ETSI's standardization of MEC has occurred independently of NFV and SDN working groups, these groups have worked increasingly closer together to ease integration of various architectures. Some of the main areas of cooperation relate to the management and orchestration of interfaces so that when deploying an MEC app on an NFV infrastructure, the relevant information is shared across the system. This has greatly simplified MEC platform deployments on existing NFV infrastructures and given operators the confidence that the shift to software-based infrastructures will not cause fragmentation that makes networks hard to manage and operate.

Although SDN has gained momentum following successful automating configuration and maintenance processes of fixed transport networks, it is still in its infancy when it comes to mobile networks. The potential of using SDN's automation capabilities in the dynamic mobile access network is considerable and MEC can be the catalyst for exploring what is possible on this front.

Technology revolutions require organizations to evolve to take full advantage of what is possible. MEC, NFV, and SDN are no exceptions and represent key elements of the transformation the operators are undergoing. For many years, operators split technology organizations based on access (radio or fixed), transmission, core and services (voice, data, and video), and IT for operations and business support systems (BSS). This approach once worked because the technology and functions performed in each group were well-defined and contained, with clear and simple interfaces between them. Each of the vertical organizations had full responsibility for engineering and deployment of the technical components, from the functional layer down to the physical. However, with MEC, NFV, and SDN emerging, operators must move to a more horizontal organizational model.

As NFV infrastructure grows across all physical technical locations, from core sites to central offices and access aggregation sites, and disaggregates network hardware from its software functions, there is a need to create an organization responsible for the end-to-end engineering, planning, and support of this new common physical layer.

The same applies for end-to-end SDN strategies, with adoption of automation tools providing immediate operational benefits. An example is reduced operational complexity for establishing network connectivity across domains. This is a must when it comes to ensuring continued evolution of operator infrastructure and disaggregation of current major block functions, like radio access and packet core.

These are all important considerations when exploring MEC because it provides a new control point that can only be leveraged most effectively by evolved organizational structures built specifically for this new world. The MEC platform enables interaction between elements of the network that relate to both radio and application performance, functions that have traditionally been addressed by separate access and core teams.

An MEC platform relies on the availability of an NFV infrastructure in the locations where it will be deployed. To maximize MEC's potential, supporting teams should bring access, packet core, and application services skills. MEC provides the opportunity to optimize some existing functions, like cell-aware traffic management, and build new services like locally hosted, low-latency applications. This is applicable to existing LTE networks and must be accommodated in future 5G networks. In short, planning an organizational structure that can best leverage the potential of MEC on LTE networks by exploring network deaggregation and experimenting with or launching new low-latency applications, is an investment in the future 5G services.

5.7 The Theory Competition

MEC is not the only initiative working on standards or reference architectures to bring some order on telecom edge initiatives. There are three groups that are worth noting and, in one way or another, related to MEC. Some operators have aligned with one or another, and they have an impact on the MEC-playing field.

- **OpenFog Consortium**, a nonprofit group founded in November 2015, and its members work on "fog computing," which adds a hierarchy of compute, storage, networking, and control functions between the cloud and endpoint devices and between gateways and devices. The nonprofit says it's different from MEC because it covers all the layers between the edge and the cloud, while MEC only covers the edge and not the cloud. OpenFog is also working with ETSI to develop fog-enabled mobile edge computing applications and technologies. The two groups signed a memorandum of understanding (MOU) stating that they will share work related to global standards development for fog-enabled MEC technologies including 5G, IoT, and other data-dense applications.
- The **Linux Foundation formed the Akraino Edge Stack Project** back in February 2018 to create an open source software stack to improve the state of edge cloud infrastructure for carrier, provider, and IoT networks. Akraino Edge Stack will offer users new levels of flexibility to scale edge cloud services quickly, to maximize the applications or subscribers supported on each server, and to help ensure the reliability of systems that must be up at all times. At this time, Akraino remains an independent initiate from ETSI's MEC.
- The **Open Networking Foundation** (ONF)'s Central Office Re-architected as a Data Center (CORD) project wasn't originally targeted at edge computing, but the ONF now realizes that its code is very relevant to the edge. In fact, the ONF stated that CORD is becoming the de facto platform of choice for edge computing because CORD can run in a tower, a car, a drone, or anywhere. Unlike ETSI's MEC ISG, which is working with OpenFog to coordinate on specs and make them interoperable, ONF said it isn't coordinating with any MEC standards group because it is pushing an open source approach.

5.8 Deep Dive on Disaggregation

Disaggregation is a group of ideas meant to separate traditionally bonded functions to achieve economies of scale, scalability, and more valuable services while combining smaller elements together in a smarter way. Network disaggregation plays an important role in how MEC will come to prominence in mobile networks and is present in many operator strategies. There are several major trends emerging:

- **Software from hardware disaggregation** is getting outsized attention because it represents a stark departure from the days of proprietary technology that operators deployed as monolithic systems or small appliances with the hardware and software functions provided by the same vendor. These included COTS routers, switches, firewalls, and compute servers bundled inside a cabinet as a "black box." But with the advent of NFV and advancements in compute technology, it has recently been possible to port traditional network functions to software capable of running on virtual infrastructure in a traditional data center. This allows the operator to manage the software and hardware life cycles separately. Because the network operating system is often tied closely to network management and automation platforms, particularly when it comes to finding and repairing faults, keeping the same software platform across several generations of hardware can help reduce costs. Also, instead of each vendor providing its flavor of hardware, the same type of compute and storage hardware can be used for various vendors, thus achieving higher economies of scale.
- **Function from appliance disaggregation** takes common appliances like firewalls or load balancing and transforms them into software-only platforms. In some cases, the set of services traditionally provided by an appliance can be broken down into smaller components. For example, a firewall could be broken into network address translation, stateful packet inspection, and deep packet services components. The resulting services can be tied together through service chaining and be scaled individually depending on the amount and type

of traffic flow. The same applies to a mobile radio or packet core function but some people might still struggle with this concept because for several decades, Telcos deployed them as monolithic services and do not yet have the engineering, planning, and deployment capabilities required to address the individual components of those platforms.

- **Control plane from appliance disaggregation** has been followed closely in the SDN community as it is the original vision of the initiative. The primary advantage is the reduction of individual network device complexity and the ability to centrally configure and control the various devices as part of a single end-to-end network domain. This is a concept that resonates well when looking at the traditional mobile infrastructure because the complex mobility or control plane signaling has always been treated separately from the data and voice user planes. But because all communication is now IP, there is an easier path toward integrating the mobility control plane with the transport control plane to create a fully software defined mobile infrastructure.

One of the drawbacks of disaggregation is fragmentation, which is one of the biggest concerns among operators as they consider strategies. MEC lights a path forward, however, by providing an "aggregation" platform to combine disaggregated functions into an interoperable and scalable environment, opening up the infrastructure to new applications that were inconceivable until just recently. But MEC is not the final destination. Rather, it is the first step toward a wider infrastructure transformation away from monolithic systems and centralized services to a fully distributed software architecture. One with best-of-breed smaller software network functions chained through service orchestration systems that dynamically allocates resources in real-time when and where they are needed.

When 5G finally arrives, it will bring a new architecture that introduces yet a new level of mobile disaggregation. But there will be many technology foundations, organizations and process adjustments, and new skills required to get there. In the meantime, 5G and LTE will have to coexist for many years as businesses will not wait for ubiquitous network generation network availability to begin delivering next-generation services. Adopting an MEC strategy in existing LTE networks with the goal of moving determinedly toward 5G is a smart way to build the foundational infrastructure, processes, and organizational skills necessary for the future.

Notes

1 www.zdnet.com/article/iot-devices-will-outnumber-the-worlds-population-this-year-for-the-first-time/
2 www.mckinsey.com/industries/telecommunications/our-insights/a-future-for-mobile-operators-the-keys-to-successful-reinvention

6

The MEC Market

The Vendor's Perspective

In this chapter, we will focus on the market perspective for the multi-access edge computing (MEC) vendors. From the hardware infrastructure providers to the software and application developers, there is a big number of potential players that can leverage this growing market. We will identify the players and discuss what are the opportunities and challenges associated with each element of the MEC architecture.

6.1 MEC Opportunity for Vendors

As we have seen since the beginning of this book, MEC provides cloud computing capabilities at the edge of the network, opening up an environment for application developers and content providers to build new low-latency applications or enhance existing ones that underperform when delivered from a high-latency public cloud. The connection between the edge and the core of the operator's network is also bandwidth constrained, so edge-hosted applications can also avoid that limitation, opening up the opportunity to move intensive computing applications from the user devices to the edge of the network. By opening their radio access network (RAN) to authorized third parties, operators will enable a new ecosystem and value chain of innovative applications and services to new vertical segments on their network.

MEC also provides the opportunity for new entrants in a segment dominated by the same vendors for several years. By providing hardware and software tailored to the needs and physical or resource limitations of edge locations, new players can gain market share in a part of the operators' infrastructure that has been mainly reserved for purpose-built network equipment, and we are talking about hundreds of thousands of locations worldwide.

6.2 Who Are the Potential MEC Vendors?

Let's explore further who are the potential vendors that will benefit from the MEC ecosystem. This list is by no means exhaustive or precise but explores the potential opportunities and segments that MEC opens up.

Mobile Operators: yes, as we have seen in Chapter 5, mobile operators can also be MEC vendors. These network providers offer the source for the network edge that brings services closer to the user. By opening up their edge to third-party applications, they become a vendor of MEC capabilities. This opportunity comes in-line with the ambition some operators have expressed in terms of transforming their network into a platform where new innovative third-party applications can be easily developed and deployed.

Application Developers: these companies build apps that benefit from edge computing, such as virtual reality or augmented reality apps. Until now, these companies have been relying on the computing power and storage of end user devices and the average connectivity and latency provided by current networks. Applications have to be developed with strong fail-safe mechanisms to cope with potential degradation of network connections, which excludes mission-critical and low-latency applications from being hosted from the cloud. Many of those mission-critical applications have been developed on user devices, raising the requirement for strong computing capabilities on those devices. The evolution of smartphones and specialized edge devices have provided a growing platform for developers; however, the requirement to support multiple hardware types and operating systems makes it very expensive for the application developers. Having a common platform at the edge of the network, with the added benefit of low-latency, high-bandwidth, and radio network information creates a whole new world of possibility for application developers. The real big question is how open and ubiquitous will this environment be across carriers and geographies.

Over-the-Top (OTT) Players: they are providers whose service goes over the Internet and can be used regardless of what network provider the user has. While there is friction as OTT services pull users away from services provided by network operators (commonly voice and messaging services), there is potential for partnering and cooperation for the best possible user experience. There will always be the net neutrality argument when it comes to OTT players and network operators; however, we are now entering a new era where operators can provide a "platform" type of service on top of traditional connectivity.

Network Software Vendors: a number of smaller vendors have emerged over the last few years, primarily focused on developing software-based network applications. The development of network function virtualization (NFV) and SDN standards has created the right environment for these vendors to gain market share; however, they have strong competition from the traditional Telco vendors that have put a fierce fight to maintain their market dominance by either reducing price points on legacy technology or offering attractive promotions for Telco to transition to NFV with them. One of the challenges of NFV so far is that it has been focusing on the existing network function to migrate from monolithic to software-based, giving the advantage to existing vendors that can provide a safer transition path to their Telco customers.

MEC creates both a completely new infrastructure layer within the telecom operators' network as well as a set of network functions inexistent until now. This is new ground the network software vendors can explore and leverage their core software expertise and ability to quickly adapt to new requirements as they are not bound to any historical technological trajectory.

Telecom Equipment Vendors: these are manufacturers of network infrastructure, such as Nokia, Huawei, Ericsson, Cisco, and others. Like in every step of the Telco architecture evolution, the incumbent tier 1 vendors are the ones best positioned to be the vendors of the new technologies. Understandably, they are some of the biggest contributors to the new architecture standards – see the number of employees they have working on ETSI working groups (MEC included) – thus influencing outcomes in favor of some of their R&D efforts, or through the acquisition of start-ups and other smaller companies.

Despite their pole position when it comes to supplying new technologies to telecom operators, these equipment vendors often delay commercialization of new technologies either because they can cannibalize other revenue streams, or because the business case is still unclear. This is usually an opportunity for smaller, nimbler providers to move faster and gain market share.

IT Platform Vendors: these vendors create the computing platforms that the applications run on. Major players include, HPE, Dell, HUAWEI, and others. The move toward virtual network functions has created a new market for traditional IT hardware vendors. On top of supplying the traditional Telco IT data centers serving operations support systems (OSS) and business support systems (BSS) applications, they have started supplying compute resources for new network-centric data centers with a much higher demand for computer resources due to the nature of the highly resilient and I/O intensive requirements of network functions. In terms of architecture, these centralized network data centers do not differ much from the IT ones, so the same hardware type can be used. MEC data centers bring new challenges and requirements to compute hardware providers. Being closer to the edge implies that the footprint of the data centers is much smaller, restricted in terms of power and air-conditioning – thus the need for new high-performance, low-footprint, and more power-efficient hardware. Companies like Viavi, ADLINK, and others have been exploring the opportunity of gaining market share through customized solutions that address the requirements referred earlier.

Interestingly, chipmakers like Intel or NVIDIA are putting a lot of effort to support both the promotion of MEC within the network operator community as well as development of frameworks that simplify the path toward MEC infrastructure. Why is that? Simply because an accelerated road to MEC will open up a completely new market for them to provide the CPUs and GPUs that will power the computing resources.

System Integrators: these companies ensure that all components work together, such as hardware, software, and networking solutions. System integrators (SI) are well-known and are utilized in the more traditional IT environments. In the network operator infrastructure, this happens typically in the OSS and BSS space. When it comes to the more traditional network infrastructure, network operators have been outsourcing the SI tasks to their tier 1 network vendors in an attempt to reduce overall costs by bundling equipment and services. This tactical approach has deepened their dependency on a small number of vendors that took the opportunity to further lock-in the network operators in their ecosystem. Also, this approach has made it harder for network operators to deploy a multi-vendor environment, since the network vendors providing SI services have no incentive to deal with the interoperability complexity and prefer to sell more of their own components, for which they have created attractive price incentives. Many times, the operators have retained the SI functions in-house in an attempt to maintain an effective multi-vendor strategy. However, financial pressures on their bottom-line have pushed them to outsource and offshore those functions in order to reduce cost. This has created an opportunity for SIs with resources in low-cost countries to thrive and gain market share. This approach works when dealing with repeatable low-cost tasks, not so much when it comes to put together a new more complex architecture like MEC.

MEC network operators will deploy hundreds, if not thousands, of remote sites and will inevitably need to partner with SIs to realize the full vision of MEC. Operators cannot be bogged down in negotiations, planning, engineering, and deployment of MEC components every time a new MEC site is deployed. This is simply net their core competence thus the need for SI partners.

Many SIs have been waiting for this opportunity. The question is: are the traditional ones ready to invest and do they have the necessary skills to do it? We are no longer talking about building a few data centers with the necessary hardware, software, and OS, but a more complex environment to be deployed across a massive amount of physical locations. Network operators should be looking for SIs that developed the necessary competency in automated services and have been developing intelligent tools that allow a faster deployment and more efficient operation of the decentralized infrastructure. This is where artificial intelligence and machine learning come into play, but that's another whole new topic for some other book…

6.3 Revenue and Cost Benefits of MEC

Market research firms such as Grandview Research[1] are forecasting the edge computing market to be over $3 billion by 2025. There is definitely value to be created in this space. The fundamental questions are: which organizations can capture that value, how, and when.

Let us look at some areas where MEC will create the highest value opportunities and who are the main vendors to benefit from it:

Network Efficiencies: for network operators, where mobile, fixed, cable, and other Internet infrastructure providers can be included, the most visible benefit will come from cost efficiencies. This is typically the approach they take since they are still very much focused on providing voice and data connectivity services. The exponential growth of data connectivity keeps pushing them toward more efficient, or cheap, infrastructure solutions.

Combined with the existing NFV/SDN work being done by mobile operators to optimize networks and reduce CAPEX, MEC can be another tool in their toolbox. It could in theory lower the bandwidth usage on their backend transport networks and the associated cost, making this one of the first angles they have been looking at MEC.

The vendors most likely to benefit from these revenue streams are the traditional network vendors as they can bundle these network efficiency services with the traditional RAN, transport, or core components. Have a look at the initial MEC use cases and what those vendors have been talking about when they refer to MEC and you will understand it.

Real Estate: the simplest business model to understand related to edge computing is simply space. Tower operators, mobile network operators, and cell site owners can rely on the collocation model and charge others to locate at certain edge locations.

This might be a surprising new revenue stream when talking about MEC and can be a more interesting one for tower operators rather than mobile operators, who are seeking higher profitable services. On the other end, there can also be the opportunity for cloud providers, like Amazon, Google, or Microsoft Azure, to start gaining share at edge real estate and provide hosting services for both mobile network operators as well as other verticals.

Unlike public clouds that rely on big centralized data centers and always look for locations with low square meter (or foot) cost or potential tax advantages, edge cloud relies on high-cost beachhead property. We are talking about locations close enough to the users and those are typically in urban locations or suburban office locations, where the property cost tends to be higher.

Enterprise Services: enterprise customers have the highest appetite for high-quality connectivity services that drive business efficiency. They have been shifting some of their key business tools toward the Software-as-a-Service (SaaS) providers, who are the most well positioned to benefit from new edge capabilities. Once at the edge, cloud service providers that offer SaaS solutions will be able to extract greater value from existing offerings or be able to deliver new ones to those enterprise customers looking to digitalize their business and move more mission-critical processes to the cloud. Simply put, cloud service providers can increase margins on their existing offerings by having a premium edge option.

Although many people believe that this is the space for traditional SaaS providers to provide a premium service leveraging the low-latency nature of edge hosting, it is more probable that new players will surface providing new mission-critical functions unthinkable up to this time. This can be an opportunity for network operators themselves to escape the inevitable destiny of becoming "bit-pipes" and create offerings to their enterprise customers. If they are smart, they will look for the right partners and develop the right business model, either through joint intellectual property (IP), acquisition of skills, or even joint ventures.

Data Center Hardware: on the hardware side, as mentioned before, the opportunity is clear. Edge data centers will need to be deployed, and serviced, opening a new revenue stream for a variety of hardware manufacturers and their upstream vendors. Edge locations are constrained in terms of space, power, and air-conditioning, which will require a new, more efficient set of hardware solutions. Although it opens up the opportunity for new players, it is likely that those will come from typical Asian countries that have been dominating the hardware space in the last years.

Applications: there is undoubtedly going to be new valuable applications that vendors can develop for the edge computing environment. It's likely to be in the form of improved customer experience and brand awareness, rather than direct monetization, on existing services (e.g., high-definition streaming of content to digital billboards). Ultimately, edge computing is another tool alongside NFV, SDN, and 5G. The monetization opportunities are there, it will just be a matter of where organizations look to unlock the value and how much of it they can capture.

6.4 What Are the Main Challenges for Vendors?

6.4.1 *Building Decentralized Data Centers at the Edge of the Mobile Network*

Building IT-like infrastructure in their networks is still fairly novel for network operators, which poses a challenge when following a best-of-breed approach for their MEC infrastructure. This may result in a typical, more conservative strategy benefiting long-time vendor–supplier relationships that will favor the traditional Telco vendors and not necessarily explore the level of innovation one might expect.

There are some recent industry initiatives, like the Linux Foundation Akraino Project[2] or the Telecom Infra Project – Edge Computing Project,[3] that are a first attempt to create a framework the telecom operators can leverage to accelerate the development of edge computing data centers, without having to rely on complex traditional IT solutions that can only be delivered through established IT or network vendors. However, this will require a more cooperative approach between telecom operators and new innovative smaller vendors in order to jointly develop the new infrastructure and reap the benefits of a more multi-vendor "open source" approach.

6.4.2 *Protecting and Securing MEC*

Probably the biggest, usually unspoken, challenge is security. For many years the network operators have been deploying architectures that physically and logically separate their user data transport from the network control layers. All of the infrastructure security and processes have been built assuming this clear separation. Since the move toward NFV this has been one of the biggest concerns of network operators and they have been pushing the responsibility to solve it from an architectural and functional perspective to their vendors. With MEC opening up even further some of the critical network control mechanisms, through radio network information services (RNIS), there is a higher degree of exposure.

There are various threats and hazards that could target MEC networks or devices, and both the network operators as well as vendors should be aware and be able to mitigate them.[4] The following is a list of some of the most common or disruptive attacks that MEC architectures or devices may be vulnerable to.

6.4.2.1 Compromised Protocols One of the most consequential attacks that MEC systems are vulnerable to is the compromising of unsecured Internet protocols. If a hacker has compromised the edge system, they may be able to both read and modify any data or network traffic that travels through the MEC infrastructure. To provide low-latency services, the network operator will have to terminate the encrypted data tunnels that usually go all the way up to the core of the mobile network, thus exposing traffic to attacks within the MEC infrastructure.

Although Internet traffic is increasingly being transported using encrypting protocols, much of the remaining uses protocols that are unsecure by default, and so consideration should always be given to what may need to be secured and what potential impacts the compromising of such protocols, like SMTP (mainly used for email) and HTTP (mainly used for unsecure web browsing), would have on businesses and network operations.

6.4.2.2 Man-in-the-Middle Attacks Speaking of protocols, they, alongside certain kinds of security measures, can also be vulnerable to man-in-the-middle attacks. These kinds of attacks are when a hacker or malicious agent intercepts, relays, and potentially alters the communications of two or more parties who believe they are communicating with each other directly.

DNS protocols are particularly vulnerable to attacks such as this; however, other protocols such as poorly configured cryptographic protocols can also become vulnerable to man-in-the-middle attacks.

6.4.2.3 Loss of Policy Enforcement The loss of policy enforcement functions such as VPN termination, IP whitelisting, or MPLS/VLAN labeling could also have an extremely significant impact on system and/or network integrity. Ensuring these situations are considered before deploying MEC systems can help to reduce their likelihood.

The failing of edge devices relied upon for enforcing security measures is something all those wishing to invest in MEC architectures will need to understand and know how to react to. If these measures fail, the hacker compromising your system would potentially have access to all the data coming from the vulnerable edge device or devices.

6.4.2.4 Loss of Data The most obvious risk from inadequate security and protective measures is the loss of data to those who may wish to intercept and steal it. Not only is personal and sensitive data at risk of interception, but also the metadata generated by edge devices detailing the type and source of data from connected edge devices.

Details such as what services and applications a user accesses, the identity of who is connecting to the network, and all the details that would be available through other unencrypted data such as email content and recipients could all be accessed by determined hackers with the right resources and know-how.

Vendors play a key role and will have to be prepared to address security concerns and provide the necessary mechanisms for telecom operators to monitor and enforce the necessary security, both on infrastructure and data transport levels.

6.4.3 Developing a Cooperative and Healthy Ecosystem

MEC opens up a new ecosystem for vendors and partnerships will be essential for the healthy development of this market. There is no one vendor that has a complete suite of solutions that serves the MEC needs. It is always possible for vendor consolidation to happen sooner than later but that might be limiting the pace of innovation and development of truly differentiating capabilities. This is a challenge for both the vendors and the network operators: there is a chance to open up the network infrastructure to new ways of building, operating, and providing network services, and it is up to all industry parties to leverage it.

6.5 Hold on ... What If the Opportunity Goes beyond Telcos?

Ever heard of private LTE networks? According to Qualcomm:

> Today's progressive enterprises, in virtually all industrial sectors, are pursuing software-driven operating models using analytics, automation and machine communications to improve productivity. These processes are underpinned by new-wave wireless

> networking solutions that offer scalable control and extreme reliability in very dense, machine-oriented environments. From Industry 4.0 factory-floor automation, to control of autonomous trucks in open cast mines, to electricity distribution grids, to logistics and warehousing, to venue services, and many more use cases, wireless networks are essential to real-time process automation and can unleash phenomenal productivity benefits.[5]

Do these requirements sound familiar? Very much in-line with what MEC purposes to provide in terms of capabilities at the edge of the network.

6.5.1 What Are Private LTE Networks?

Some industries are looking for more robust communication solutions that can ensure the high-performance levels required for their mission-critical systems. The current public LTE networks are not able to provide the service levels they require; thus the search for dedicated solutions that can serve the raising demand for production-critical automation and mobility needs.

Enterprise organizations that control their own networking environment can more easily optimize it for their own purposes. The availability of open access spectrum, like the 3.5-GHz band in the U.S. and the 5-GHz unlicensed band globally, makes it possible for almost any organization to deploy and operate private LTE networks (Figure 6.1).

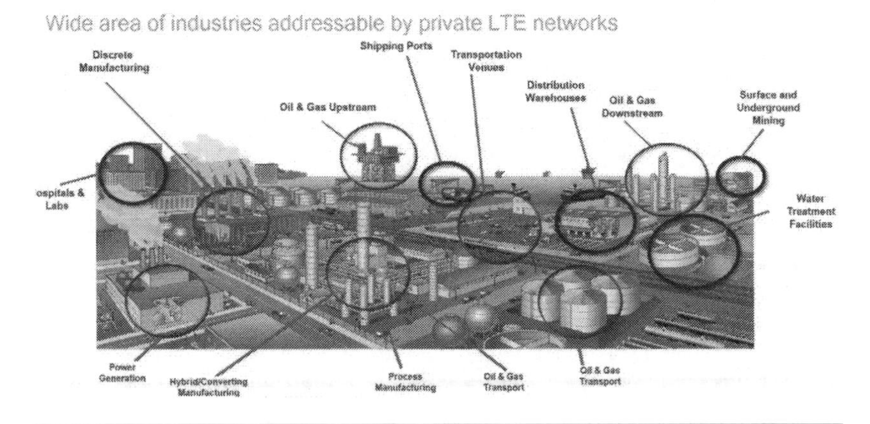

Figure 6.1 Examples of industries considering private LTE networks. (Source: Qualcomm.)

6.5.2 Is MEC Being Considered for Private LTE Networks?

Not necessarily. The motivation behind private LTE networks is for enterprises to build and operate their own dedicated network completely segregated from the public networks. This allows them to control which devices access the network and ensure the necessary quality of experience (QoE). Most of the times the enterprises are assuming building localized private networks on a per location basis, resulting in multiple instances across diverse geographical locations.

By adopting an MEC architecture, the same objectives can be achieved with a significant simplification of the overall architecture and cost reduction. Instead of building multiple segregated mobile core infrastructure per campus/building location, the enterprise can build a distributed MEC platform, per campus/building, and a single centralized core (e.g. one serving all locations in a country or state). This greatly simplifies operations across multiple locations and provides seamless access across the various locations, whilst enabling specific localized services provided through the on-premise MEC infrastructure.

6.5.3 What Is the Opportunity for MEC Vendors?

In a green field approach, there is higher appetitive to adopt new solutions that can provide a more efficient and future-proof solution. By engaging with enterprises that are planning to build private LTE networks, vendors can leverage both their knowledge of traditional networks as well as their understanding of mission-critical applications and its network requirements. The fact that this type of customers have a clear idea of which applications they want to run at the edge and the connectivity requirements greatly simplify the business case conversation and creates a good initial environment to effectively leverage the MEC capabilities. In summary, there is no better opportunity than a green field approach with clear business requirements to kick-start MEC deployments from scratch, rather than working on a major transformation of existing (complex) infrastructures.

Notes

1 www.grandviewresearch.com/press-release/global-edge-computing-market
2 www.lfedge.org/projects/akraino/
3 https://telecominfraproject.com/edge-computing/
4 www.lanner-america.com/blog/multi-access-edge-computing-part-2-security-challenges-protecting-securing-mec/
5 www.qualcomm.com/media/documents/files/private-lte-networks.pdf

7

The MEC Market

The Over-the-Top Player's Perspective

In this chapter, we discuss various ways in which the over-the-top (OTT) actors may approach the multi-access edge computing (MEC) marketplace and the reasons to choose one approach over another.

7.1 The OTT Approach to the Edge

The potential interaction between communication services providers (CSPs) and OTT applications and cloud providers has already been highlighted in Chapter 3. With MEC, the CSPs are in a position to offer its services to the OTT service providers. In this chapter, we examine this issue from the OTT provider's perspective – why and how would they take advantage of the MEC services that the CSPs have to offer.

The "why" is rather straightforward. With few exceptions, cloud service providers and cloud application providers do not have access to physical locations that place them within physical proximity of their users. As the requirements of emerging applications drive the need for proximity to the end user, the natural proximity of the CSPs becomes the foundation for a business relationship between the CSPs and OTT actors.

In the rest of this chapter, we shall concentrate on the various ways that the OTT actors may approach the MEC marketplace and the reasons to choose one approach over another. We start by examining the kinds of actors that would be OTT players from a CSP perspective. Generally speaking, by *over-the-top*, we mean any entity whose primary interaction with the CSP's access network is that of an IP pipe. We may partition such actors into three broad categories:

- Cloud service providers, for example, Microsoft Azure, Amazon Web Services (AWS), and Tencent
- Cloud-based Software-as-a-Service (SaaS) providers, which may include SaaS services provided by the cloud providers, but also includes other players, for example, Oracle
- Cloud-based application providers, which may be small application developers, but also developers or large-scale distributed applications, for example, for gaming or vehicular automation. In this case, it is often useful to distinguish between small and large application providers. In particular, these may differ in the following aspects:
    - Large application providers have the capability and scale to enter into direct agreements with a large number of CSPs; small application providers may not.
    - Large application providers may thus be able to request support of custom HW needed to run their applications. Small application providers (if they are cloud application providers) must assume generic, abstracted resources only.
    - Large application providers may have the scale to support their own HW deployments and operationalize application deployments over large distributed clouds; small providers do not.

In Chapter 3, we examined the business models that the CSP's may use to address the MEC market and, following the analysis in Ref. [76], identified six of them. Of these, edge hosting/co-location and edge XaaS are likely to be the two most common CSP business models encountered by the OTT players. The other three models listed in Ref. [76] involve significant participation by the CSP in the application delivery process and therefore cannot be considered "over-the-top"; while the sixth model added by us focuses on optimization of the CSPs internal OPEX. As such, we concentrate on the edge hosting/co-location and edge XaaS, and examine the advantages and drawbacks of each of these MEC models for the three types of OTT players we have identified.

7.2 Edge Hosting/Co-Location

We start with the edge hosting/co-location approach. In this business model, the OTT actor brings its own full solution, including the necessary hardware infrastructure, to the MEC site. The CSP has to provide space (usually on the rack) and power. It also typically provides ability to access traffic on its access network; however, this is not a universal need. Likewise, transport to the Internet (and thus to the OTT's main cloud) is typically, but not necessarily, provided by the CSP.

From an OTT's provider's point of view, this enables them to retain complete control over this portion of the distributed cloud. It is also the only way for them to provide an Infrastructure-as-a-Service (IaaS) offering at the edge. As such, this model may be quite attractive to the large cloud providers – they want the control and it enables them to extend the full range of their offerings to the edge. In those cases where the application requires specialized hardware or operating system feature (e.g., a real-time OS) not typically present in cloud deployments, this makes it possible to deploy such systems at the edge.

However, this approach has a number of drawbacks. The OTT actor has to fully manage their edge cloud. This typically includes maintenance of the hardware, which means processes and organizations have to be established to do so over a large number of locations with little infrastructure in each one. For a cloud provider, this is in contrast to the typical cloud data center deployments that concentrate a large amount of hardware infrastructure in a small number of locations. For other actors, this means a return to owning and managing a large amount of hardware (and, in particular, significant associated CAPEX) – precisely what the emergence of cloud eliminated. In the case of smaller size players, this is simply not possible.

Another drawback is the provisioning of SLAs (i.e., service agreements) that go beyond the typical best-effort service of cloud deployments. CSPs offer services which include critical infrastructure and guaranteed-QoS type services. As such, they design their networks and their internal clouds to be able to meet the requirements of such stringent SLAs. In the co-location market model, an OTT player does not have access to any of these and must design their own cloud to meet such stringent SLAs if this is necessary.

As such, the co-location model is the most attractive in two particular OTT cases. The first is the large (web-scale) cloud provider that needs full control of the edge clouds and wishes to offer IaaS at the edge as a service. The second is a large-scale application provider (e.g., a continent-sized vehicular automation provider) that requires custom infrastructure and OS. In both cases, the scale of the OTT actor makes it feasible and economically reasonable to deal with the maintenance and operation issues that come with running their own edge cloud.

7.3 XaaS

In the XaaS model, the CSP owns and manages the cloud infrastructure across its MEC locations and offers this infrastructure as a service to the various OTT providers. Because the physical infrastructure is owned by the CSP, the CSP is responsible for the operation and maintenance of this infrastructure. The OTT actor also does not carry any CAPEX burden. This makes the approach attractive to OTT actors that are already entirely in the cloud and those who simply do not have the scale to operate infrastructure at a global level.

Another advantage of a CSP-owned and -operated infrastructure is the ability to deliver SLA-differentiated services. Given the needs of their own internal clouds, CSPs have internal know-how on how to design and deploy edge cloud infrastructure that can meet requirements such as those of critical infrastructure, public safety, and high availability (e.g., for industrial automation). These SLAs are not typical in existing cloud deployments. Yet, with CSP-operated infrastructure, applications can simply request such SLAs as part of the service (presumably, a cost is associated with these).

The issue of application deployment is also an important one. Understanding when and where to deploy edge instances of application components is a highly complicated task and getting it wrong can mean significant impact on both user experience and cost. While some application providers will do this themselves, many, especially smaller ones, lack the scale and know-how to do so. With an XaaS model, the CSP can take over providing this as a service offered to the applications running on its edge cloud (e.g., as part of an SLA agreement with the OTT application provider). Other similar services can also be provided by the CSP. These may include:

- Distribution of data across the distributed edge storage.
- Access to a well-designed distributed edge database.
- Security-related services (for edge access, authorization, etc.).
- Other value-added services.

While some of these can also be made available in the edge hosting market model, the XaaS approach makes offering and consuming these quite a bit more straightforward.

There are disadvantages to the XaaS market model as well. The most significant one is that the OTT player cannot offer IaaS services in this model (IaaS is not nestable). Fortunately, other types of cloud resource–as-a-Service models (Platform-as-a-Service [PaaS], Function-as-a-Service [FaaS]) are possible and so this limitation is mostly applicable to the very large OTT players that traditionally offer IaaS.

A second limitation is the fact that the cloud resources are limited to what the CSP has deployed. For example, if the CSP has very limited GPU capability in the edge clouds, GPU-intensive applications may not be able to properly operate in such clouds.

We summarize our discussion in Figure 7.1, which attempts to capture the aspects of the MEC Market Models and the various OTT actor types side-by-side. On the MEC Market Model side, a white color is used to indicate full support of a feature, a black is used to indicate lack of support – or major issues with providing it.

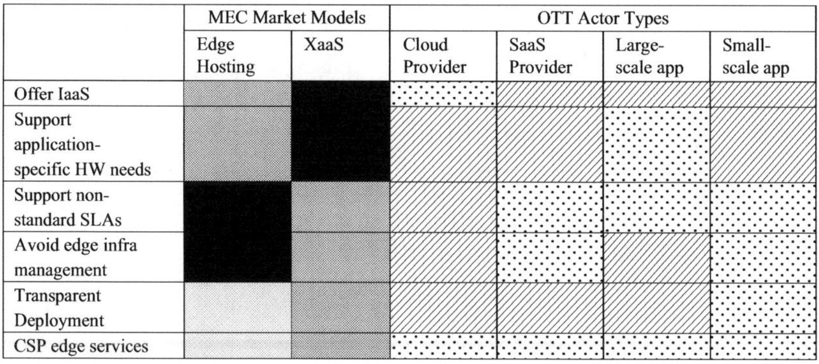

	MEC Market Models		OTT Actor Types			
	Edge Hosting	XaaS	Cloud Provider	SaaS Provider	Large-scale app	Small-scale app
Offer IaaS						
Support application-specific HW needs						
Support non-standard SLAs						
Avoid edge infra management						
Transparent Deployment						
CSP edge services						

Figure 7.1 Visualizing the OTT View [44].

Finally, gray is used to indicate that support may be limited. For the OTT Actor Type side, dots are used to indicate high need for a feature, while stripes is used to indicate low-need. The extent to which white on the left and dots on the right match – and black on the left and dots on the right do not – is a good indicator of how appropriate a particular MEC market model is to each particular type of OTT player.

8

The MEC Market

The New Verticals' Perspective

The present chapter describes the main benefits of multi-access edge computing (MEC) from the point of view of vertical market segments. We will first describe the heterogeneous ecosystem of verticals, and then also provide a more detailed case study, as a meaningful example for the automotive sector.

8.1 New Players in the 5G Equation

When it comes to mobile phones, and referring mainly to 2G voice communication, there is a commonly established consensus in considering mobile operators as the main players in the communication market, at least traditionally.

Nevertheless, with the advent of mobile data and broadband connections (thus 3G and 4G), new stakeholders have entered this game, the so-called over-the-top (OTT) players. These companies, like Google, Facebook, Netflix, Amazon, etc., generally provide services across IP networks, and by default consider the operator network as a simple "bit pipe". So, even if operators are still the main infrastructure owners, OTT players are exploiting their network to deliver added-values services (e.g., video content and streaming services) to end customers. So, in terms of global revenues, we can also say that these players conquered in few years a major part of the market shares, essentially starting from zero. Moreover, this trend doesn't seem to stop: in fact, according to forecasts, global OTT revenues are expected to double in 2023 (with respect to 2018) [78].

More recently, other new players have also been joining the party: the so-called vertical market segments. These companies typically

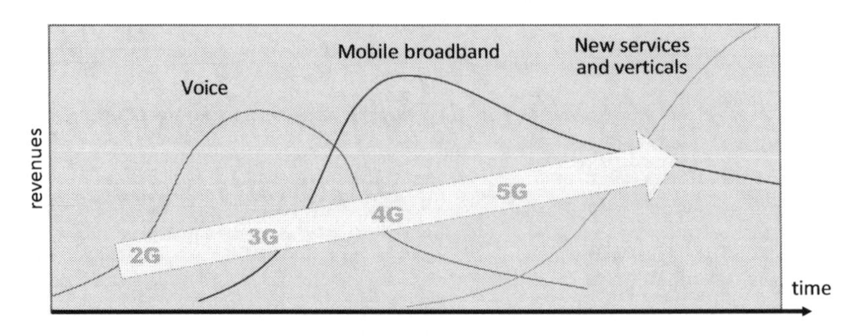

Figure 8.1 Waves of services and their revenue streams correlated with the network evolution.

come from various and heterogeneous business areas, and drive new use cases for the communication systems; for example:

- automotive and cooperative vehicles,
- virtual reality/augmented reality,
- Internet of Things (IoT) (sensors, fog nodes, etc.),
- robotics and factories of the future,
- eHealth and mHealth,
- media and entertainment vertical,
- applications for the energy industry.

These typically also correspond to 5G use cases, where industry verticals are indeed bringing new technical requirements, and practically driving the evolution of communication systems toward 5G (and beyond). Most of the market forecasts are also emphasizing that revenues coming from vertical streams will increase in the future, and potentially overcompensate the decrease of value coming from voice and traditional data services. Figure 8.1 shows in a qualitative way the described trend, thus associating waves of services and their revenue streams with the network evolution (for more details about the so-called "waves", the interested reader can have a look at Ref. [77]).

As an important clarification, new revenue streams are not expected to be captured mainly by operators. On the contrary, the ecosystem is becoming more complex and imposing different business models, including collaboration, revenue sharing, different level of partnerships, etc. As a meaningful example of new market value generated by verticals in 5G, we can consider IoT: according to GSMA forecasts [79], the total ecosystem IoT revenue in 2024 will be around 4,300 billion dollars, while only 86 of them are expected to be captured by operators in this domain. Similarly, we can also expect that other 5G services will be characterized by this value fragmentation.

8.1.1 The Role of Verticals in 5G Systems

Taking again the IoT space as an example of driver for 5G system, one of the commonly recognized limitations of 4G systems was that it was not designed (e.g., in terms of signaling) to efficiently scale for enabling massive machine-type communications (produced by the incredibly huge number of connected devices expected in the next years). This trend is in fact universally considered as one of the main differentiating factors between 4G and 5G systems. We are talking about a myriad of sensors and low-cost, low-power devices, but also any kind of connected devices collecting data and communicating with the cloud infrastructure.[1]

Another example of verticals influencing the advent of 5G is provided by the automotive sector, and especially industrial automation and robotics, where ultrareliable and low-latency communication is critical for the service delivery. This is also the case of virtual reality/augmented reality use cases, which can be applied to many scenarios (from tele-operated driving and assistance, to gaming, and advanced social network applications…).

In summary, each new vertical imposes new and heterogeneous requirements to the forthcoming communication systems. In addition, for most of these use cases, edge computing is a key supporting technology (as well as the radio interface), since it provides multiple benefits in terms of both low latency and data transport savings to the network. In order to better focus on MEC benefits for verticals, we need then to analyze the various industry verticals and their relation to 5G and MEC.

8.1.2 Overview of Vertical Segments and Their Impact

The following list of vertical market segments is not exhaustive, but provides an overview of the different industry sectors, sometimes also grouped in categories.[2] A careful reader may observe that the definition of each vertical is not an easy task, as all these sectors are characterized by a number of heterogeneous stakeholders, and often traditional companies are not only accompanied by newcomers, like small enterprises, but also by operators and big giants of the digital transformation. Moreover, one may also expect the arrival of start-ups as typical "disruptors" that may also change the market composition (as Airbnb and Uber did, respectively, for real estate and taxi markets). Nevertheless, a tentative categorization of industry sectors and companies is still a good starting point to group use cases and identify the requirements for MEC.

8.1.2.1 Smart Transportation

Sectors	Automotive, railway systems
Examples of companies	BMW, Daimler, FCA, DB
Relevant bodies and associations	5G Automotive Association (5GAA), Automotive Edge Computing Consortium (AECC)
Description	This sector is mainly driven by car makers and the ecosystem of automotive suppliers. Their requirements are derived by the need to deliver solutions for connected and automated driving cars, by exploiting 5G and MEC, that is, cloud computing at the edge of the network. The innovation in this section is relatively recent, as traditionally this ecosystem did not consider network connectivity or MEC. Only recently, associations like 5GAA and AECC are putting together stakeholders from the two worlds (automotive and communication) in order to facilitate dialogue and establish a common basis of understanding on the practical implementation of solutions for connected and automated driving cars.
Relevance of use cases to MEC	A comprehensive list of use cases can be significant for connected vehicle applications. Most of them need high performance network connectivity and edge computing capabilities. In Section 8.1.2, we will provide an extensive case study on automotive vertical.
Main benefits given by MEC	Benefits range from low latency and data transport saving, to security (related also to local access to information at the network edge).

8.1.2.2 Smart Manufacturing

Sectors	Industrial automation, cloud robotics
Examples of companies	Bosch, ABB, Siemens, Beckhoff, Audi, HMS
Relevant bodies and associations	5G Alliance for Connected Industries and Automation (5G-ACIA)
Description	This sector is mainly driven by manufacturing companies and the ecosystem of industrial automation suppliers. Their requirements are derived by the need to deliver solutions for connected and automated industries, by exploiting 5G and MEC, that is, cloud computing at the edge of the network. The innovation in this section is relatively recent, as traditionally this ecosystem did not consider network connectivity or MEC. Only recently, associations like 5G-ACIA are putting together stakeholders from the two worlds (industrial automation and communication), in order to facilitate dialogue and establish a common basis of understanding on the practical implementation of solutions in this field.
Relevance of use cases to MEC	Several use cases have been coming from the industrial automation domain, mainly showing the need to use 5G in a factory in the future. Promising application areas range from logistics for supply and inventory management, through robot and motion control applications, to operations control and localization of devices and items. In particular, the relevance to edge computing is particularly evident when these stakeholders are rethinking existing processes and introducing new processes for the transmission, handling, and calculation of production data. One example is local data centers that support critical industrial applications by way of an edge computing approach. In this case, industrial applications can be deployed locally within an edge data center to reduce latency.
Main benefits given by MEC	Benefits are mainly due to low latency, but also more in general related to the possibility shift to software-based solutions.

8.1.2.3 Entertainment and Multimedia

Sectors	Gaming, VR/AR, and content distribution network (CDN)
Examples of companies	Sony, Dolby, Akamai, Sky
Relevant bodies and associations	Virtual Reality Industry Forum (VR-IF), Virtual Reality Augmented Reality Association (VRARA), etc.
Description	This sector is composed of a broad range of stakeholders from entertainment, gaming, and innovation companies in the VR/AR ecosystem, and also video and content providers. In particular, VR/AR technologies are essentially focused on how people interact with and experience the physical world, how they are entertained, and how services are delivered to them. The continuous innovation in this sector has been imposing strict requirements on the network connectivity, and recently, also on MEC. Industry associations like VR-IF are putting together applications and content creators and distributors, with consumer electronics manufacturers, professional equipment manufacturers, and technology companies. VR-IF is not a standards development organization (SDO), but is collaborating with them for the development of standards in support of VR services and devices. Recently, in VRARA, a new 5G Industry Committee will be pursuing VR/AR focused use cases and requirements for 5G networks so as to ensure that the resulting specifications address the needs of this key industry sector. The work of this committee will be also related to MEC.
Relevance of use cases to MEC	The need is to exploit communication networks that will have to support these critical applications, delivering the required performance, for example, latency on the order of several milliseconds. Edge computing is necessary to deliver such a performance.
Main benefits given by MEC	Delay is the main issue, and it is essential to deliver real-time multimedia services to end customers. Moving from remote cloud to edge cloud is a natural evolution of the technology.

8.1.2.4 eHealth

Sectors	Remote assistance, telemedicine, connected hospitals
Examples of companies	Hospitals, medical centers, governments, start-ups (Reliant), and also big digital companies (Google, Apple, IBM)
Relevant bodies and associations	Personal Connected Health (PCH) Alliance
Description	According to Arthur D. Little, the global market for digital health will grow to more than USD 200 billion through 2020. The digitalization of health care is a wide sector, and includes cloud-based medical records to digital pills, as nearly every aspect of the medical industry is under transformation. In addition, medical processes and medical products will generate a huge amount of data in the next future. Moreover, connecting patients and doctors with all that medical information is the job of digital health companies. For these reasons, this sector is not only composed of big medical companies, but also by start-ups and new comers, driving a digital health application ecosystem (e.g., to provide users access to their medical records, medical appointment information, and prescriptions). Moreover, telemedicine is another emerging area, including many use cases (novel treatment services, high-resolution video, telepresence, augmented reality, virtual reality, etc.). This sector is also stimulating a huge market of wearable and connected devices, known as Internet of Medical Things (IOMT), consisting of a myriad of smart/mobile sensors and medical devices that support monitoring and are conceived to help advance the future of medicine. For all these applications, it is a consolidated trend also the increasing usage of technologies like blockchain or artificial intelligence (AI) applied to health care.
Relevance of use cases to MEC	A typical example of eHealth use cases is given by 5G wireless connectivity inside the hospital, enabling remote or robotic surgery, and including possibly high-quality and low-delay video transmission, or elaboration of 3D images and large image data, or again coupled with VR/AR applications and devices.
Main benefits given by MEC	Essentially, edge computing will provide low latency, and the possibility to enable VR/AR applications, remote/robotic surgery use cases. Also, adding processing and computing at the network edge. In addition, one of the obstacles to the development of the mobile health market is that information systems of most public hospitals are in low openness, low standardization, and low ability of data sharing. MEC, as a suitable standard for data exchange interoperability, could help in that perspective.

8.1.2.5 Smart Cities

Sectors	Energy efficiency, smart buildings, smart grids, tourism
Examples of companies	Schneider Electric, Cisco, Mitsubishi, Toshiba
Relevant bodies and associations	OpenFog Consortium (OFC)
Description	A lot of use cases can be included in the category of "smart cities", involving the usage of a myriad of sensors and connected devices (in this context also called fog nodes). As a consequence, a huge ecosystem of stakeholders is by definition working on this area. As a relevant industry group, the OFC was established few years ago to create an open reference architecture for fog computing, building operational models and testbeds, defining and advancing technology, educating the market and promoting business development through a thriving OpenFog ecosystem. Currently, OFC is incorporated in the Industrial Internet Consortium (IIC), which covers all IoT segments, including Energy, Healthcare, Manufacturing, Mining, Retail, and Transportation.
Relevance of use cases to MEC	If the advent of IoT is one of the drivers for 5G systems, on the other hand, edge computing is a key supporting technology for IoT and smart cities. Examples of MEC use cases supporting smart cities are: active device location tracking, security & safety, and data analytics & big data management (e.g., through massive sensor data preprocessing).
Main benefits given by MEC	Essentially, edge computing will provide low latency and the possibility to locally process a huge amount of data at the network edge, coming from a sensor and connected devices.

8.2 Benefits of MEC: A Vertical's Perspective

As we have seen in the previous section, all industry verticals impose different and multiple KPIs on to the MEC-enabled 5G systems. From a communication point of view, all these requirements result in a heterogeneous network, for example, managed through network slicing. In fact, network slices are the essential tool that permits the management and satisfaction of heterogeneous KPIs through flexible allocation of dedicated (radio and core) network resources. On the other hand, the MEC system is the application end point for all these identified use cases, and it is expected to contribute to the E2E performances (i.e., from an application client running on terminals/devices to the edge server instance running on an MEC host).

In the following sections, we provide an overview of the main MEC benefits in common to all segments, always from a "vertical" point of view (i.e., not necessarily motivated by specific operator-driven perspectives): from the need for performance improvement, to the unlock from operators ownership and control, and the adoption of standards as a means to guarantee interoperable IT solutions.

8.2.1 Performance Improvement

Network KPIs are seen by verticals as a priority from a service level point of view. In particular, E2E latency is often one of the most commonly recognized KPIs where MEC provides huge and evident benefits.[3] Nevertheless, the actual MEC performance should be compared against different deployment options (shown in Figure 8.2), and thus the MEC gains depend on the specific vertical/use case considered (left-hand side of the figure).

In the context of delay, especially when latency constraints are critical (e.g., few milliseconds), a primary role is played by the radio access. In this perspective, 5G connectivity, and the introduction of ultra-reliable low-latency communications (URLLC) is a key priority for vertical industries, especially for some delay- and mission-critical services, for example, automotive, industrial automation, and VR/AR.

Overall, since the target is to ensure end-to-end performance, both the network infrastructure (which is falling into the 3GPP domain) and the MEC infrastructure play a key role for the satisfaction of those E2E requirements. From a business point of view, both 5G network and MEC systems are expected to be part of the operator domain, who could be most likely hosting the whole infrastructure, and offer (as a service) to third parties, thus planning to establish a deal with the respective verticals. As a consequence, not only the 5G network but in principle also the MEC system have an impact on the service level requirements (SLR) and service level agreement (SLA) between the vertical and the MNO.

8.2.2 Unlock from Operators

Even if a collaborative approach between verticals and operators had emerged in these years (i.e., driving many requirements for the development of 5G systems), more recently, parallel initiatives

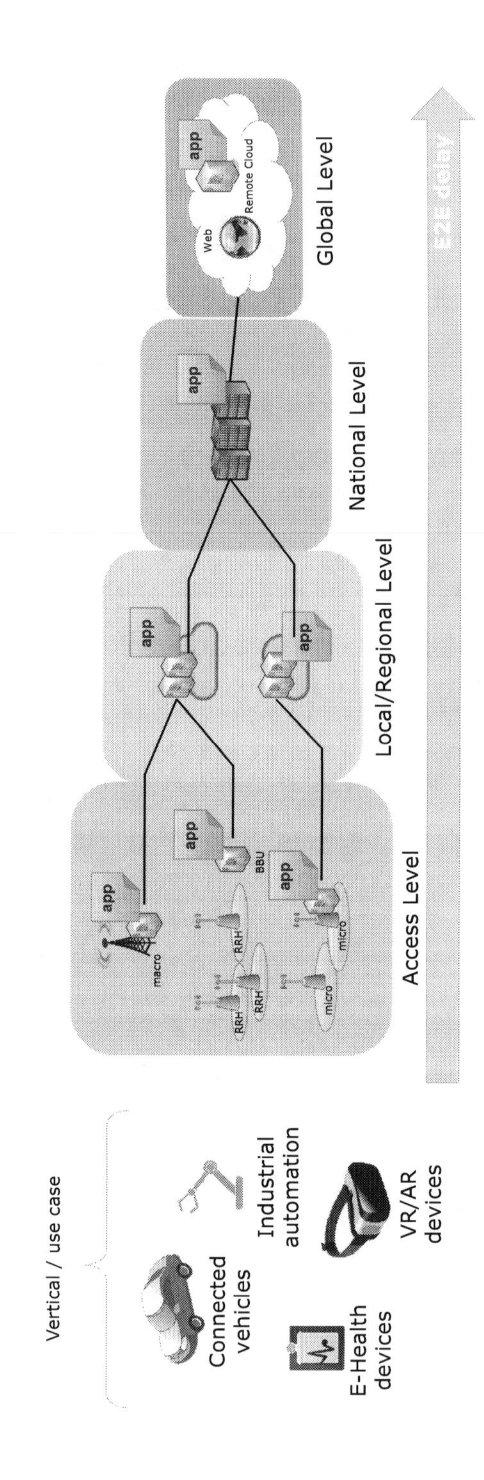

Figure 8.2 Different MEC deployment options and related E2E delay performance.

from vertical segments have also revealed their impellent need to deliver services in contrast to (or at least independently from) operators.

The most evident example is provided by the industrial automation domain, where in contrast to a network that offers mobile network services to the general public, some identified scenarios foresee a 5G nonpublic network (NPN; also sometimes called a private network) that provides 5G network services to a clearly defined user organization or group of organizations [80]. In some of these scenarios, the NPN and the public network share part of the radio access network (or also the control plane), and hence some level of sharing can be implemented, for example, through network slicing. In other cases of isolated deployment, instead, the NPN can be deployed as a standalone NPN. Another aspect, showing this dual/controversial relation between verticals and the operator, is the usage of assets traditionally owned only by MNOs: the spectrum. In some countries, the regulator allows verticals to request a spectrum for specific usages (e.g., for industrial automation). As a consequence, it is clear how industrial automation stakeholders are in principle investigating many options, to be able to innovate their production plants by exploiting advance communication networks, while preserving their needs of privacy, data ownership, etc.

From an edge computing perspective, the vertical often prefers to consider cloud solutions as a mix of infrastructures: since MEC applications are in principle software instances that can be conveniently placed in many locations, some critical workloads (or confidential databases) are preferably running (stored) in private clouds, while other instances can be deployed in the MNO infrastructure. In these cases, the main concern is of course the security, since the vertical is primarily interested in preserving the relation with the end customer, who expresses his consent to share personal data, or any kind of confidential information, for the delivery of a specific service.

In summary, although it is not totally clear, at the present moment, which kind of technical solution and level of collaboration (and related business model) will be adopted at the end by verticals, as a matter of fact MEC is also seen today as a tool, offering them many options and degrees of freedom for deploying their services.

8.2.3 MEC as Enabler of IT Interoperability in 5G

Another important aspect of E2E service deployment is the awareness that industry vertical players need to collaborate with an articulated ecosystem of technology providers (from mobile operators, to infrastructure providers, chip manufacturers, system integrators, application providers, etc.). In this complex environment, composed by multiple stakeholders, an E2E solution needs a proper communication interface between the different modules implemented (especially when, e.g., different application instances are developed by different companies and need to exchange data, or produce/consume common datasets, e.g., from sensors and cameras). In particular, MEC as the only international standard available for edge computing is offering an evident means to enable interoperability for data exchange. In this context, the standard APIs specified by ETSI ISG MEC are in fact defining data types and formats, and suitable messages (through RESTful commands) that help application developers to design complete solutions, with the advantage to enable portability and reduce time-to-market for service creation, operation, and management.

8.3 5G Verticals: Business Aspects

We have seen that all the described vertical segments are driving the traffic explosion and are imposing new requirements for the introduction of 5G systems. Figure 8.3 shows, in particular, the expected number of connections, differentiated per traffic type.

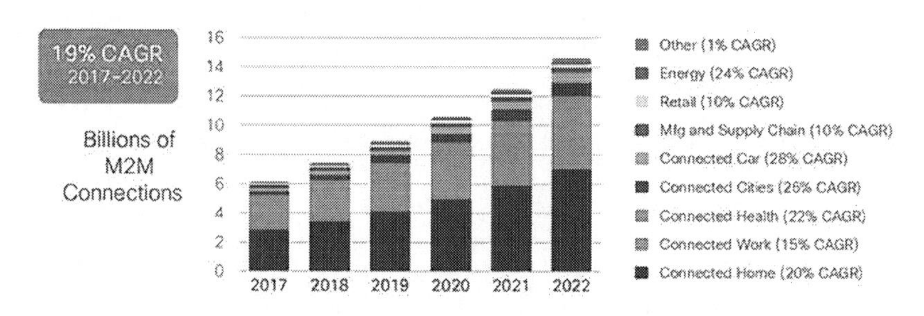

Figure 8.3 Global M2M connection growth by industries. (Source: Cisco [82].)

Nevertheless, for each connection (related to a specific vertical segment) in order to correctly identify the addressable markets for the different vertical segments, it is worth considering specific forecasts, related expected revenues. In fact, the various vertical segments address different markets (each one with its own characteristics in terms of revenue potential and also expected year of service launch). Here is a brief overview of the different total addressable markets (TAMs) for each identified vertical:

- **Automotive**: according to Riot Research [81], the total size of the automotive connectivity market will grow from $7bn in 2017 to **$29.2bn in 2023**. This is of course including all revenue sources, that is, from cellular telematics hardware, to mobile network operator connectivity revenues, revenues derived from reselling connected car data, V2X hardware sales, and connected car services subscriptions. In particular, the annual Cellular Hardware Revenue will grow to $1.98bn (related to a total volume of 427 million connected cars), while a bigger revenue portion is expected to come from Annual Connected Car Data Revenue ($13.7bn).
    - From an MEC perspective, the actual market potential depends on the MEC hosts deployment scale, for example, whether the MEC hosts will be deployed co-located with roadside units (RSUs) or cellular radio base stations, or will again be associated to a cluster of base stations. This choice will depend on the infrastructure owner (e.g., mainly, the 5G mobile operator).
- **Industrial automation**: according to [83], the global factory automation market size is expected to reach **$368bn in 2025** (from $191bn million in 2017), growing at a CAGR of 8.8% from 2018 to 2025. It comprises the use of computers, robots, control systems, and information technologies to handle industrial processes (Figure 8.4).

  According to other estimations on the smart manufacturing market [84], global revenues are expected to "reach approximately **USD 479.01 billion in 2023**, growing at a CAGR of slightly above 15.4% between 2018 and 2023. The adoption of smart manufacturing technologies by global players will

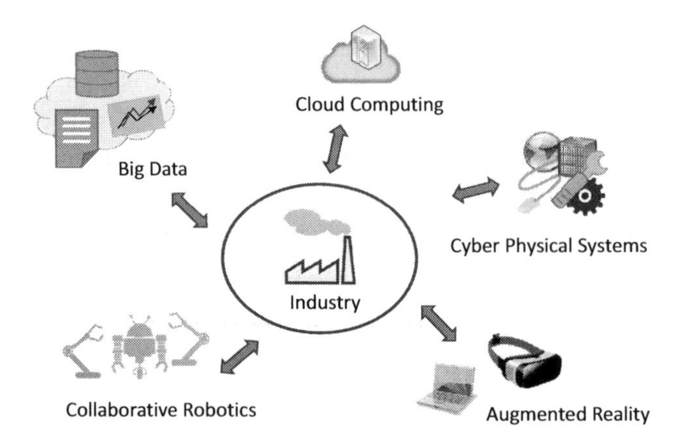

Figure 8.4 Different components of the factory automation market.

provide opportunities for automating operations and use data analytics to improve manufacturing performance."

- Here, estimations are very heterogeneous, as the market is potentially quite large. From an MEC perspective, the actual market potential depends on the MEC hosts deployment scale, for example, whether MEC hosts will be deployed co-located with a small cell, or a production plant (and then associated to a cluster of base stations). This choice will depend on the infrastructure owner (e.g., mainly, the 5G mobile operator, or the OT player, in case of stand-alone NPN deployments, or both, in case of network sharing).

- **VR/AR/multimedia**: market forecasts in this sector are very heterogeneous, and ranging from more optimistic expectations (Statista.com predicts that the AR/VR industry will be valued at **$209bn by 2022**), to more conservative ones (Goldman Sachs Research predicts that the AR/VR industry will be worth **$80bn by 2025**). This discrepancy is understandable as its growth is dependent on the growth of complementary technologies such as 5G (and also on the actual penetration of proper devices, e.g., AR glasses).
  - From an MEC perspective, the actual market potential depends on the MEC hosts deployment scale in 5G networks, for example, whether the MEC hosts will be deployed co-located with cellular radio base stations,

or will again be associated to a cluster of base stations. This choice will depend on the infrastructure owner (e.g., mainly, the 5G mobile operator). In addition to that, other players may join in the future, for example, cloud providers, which are today offering cloud services without the need of MEC. If these players will decide to expand their footprint at the edge of the network, of course, an estimation of the TAM will also have to take into account this aspect (e.g., additional MEC servers owned by cloud providers).

- **eHealth**: According to Arthur D. Little, the global market for digital health will grow to more than $200bn through 2020. Other estimates [85] are talking about a growth of the eHealth market up to **$132.35bn** revenues by **2023** (from $47.60bn in 2018), at a CAGR of 22.7% during the forecast period.
  - Also, in this case, eHealth services are expected to be provided through 5G systems. Thus, from an MEC perspective, the actual market potential depends on the MEC hosts deployment scale in 5G networks, for example, whether the MEC hosts will be deployed co-located with cellular radio base stations, or again associated to a cluster of base stations. As previously clarified (please refer to Figure 8.2), this deployment choice can differ from other verticals/use cases, and in any case, will depend on the infrastructure owner (e.g., mainly, the 5G mobile operator). Again, additional revenue sources should be considered, in case, for example, connected hospitals host some dedicated cloud infrastructure.
- **IoT/Fog/Smart Cities**: IoT Analytics predicts that the global market for IoT (i.e., end user spending on IoT solutions) is expected to grow 37% from 2017 to $151bn in 2018 [86]. Due to the market acceleration factors, they revised estimates upward, and it is now expected that the total market will reach **$1.567bn by 2025**.
  - Here the market is more fragmented (and heterogeneous), as it potentially includes all the revenue streams coming from the IoT world. When it comes to an estimation of

the MEC market potential for this sector, we may reasonably assume that most of the IP traffic will be captured by 5G mobile networks and Wi-Fi indoor/outdoor hotspots, the actual market potential (from an MEC perspective) depends not only on operator choices about 5G (and the related MEC hosts deployment scale) but also on enterprise and retail scenarios, including airports, shopping malls, etc., where MEC hosts will be likely deployed as co-located with Wi-Fi access points. This choice will depend on the infrastructure owner (e.g., mainly, the enterprise or building owner or, again, the Wi-Fi operator).

From these estimations, it is clear how MEC is a key supporting technology for all the described sectors. Nevertheless, it is not easy to define the exact potential for MEC, as the actual deployment would mainly depend on (i) the MEC hosts deployment scale in 5G networks, based on operators' choices, (ii) the related deals with industry verticals, and (iii) the evolution of terminals and devices. In particular, the second item is described in the following section.

8.3.1 Cost Aspects

MEC deployment models and associated costs will be described in detail in Chapter 9. In this section, we will introduce some aspects specifically from an industry vertical perspective. In particular, the costs associated with MEC deployment depend on the actual business model (e.g., more or less collaborative with the 5G operators). For this reason, it is not easy to predict which kind of model will be adopted for MEC. Nevertheless, since it is unlikely that the entire infrastructure will be owned by the industry vertical (as this would be a very extreme case, and possibly considered only for particular use cases), the most reasonable hypothesis is to consider the different levels of cloud computing offerings that can be proposed typically by the MNO: Infrastructure-as-a-Service (*IaaS*), Platform-as-a-Service (*PaaS*), and Software-as-a-Service (*SaaS*). In particular, when it comes to edge computing, industry verticals may be interested in the first two models offered by the operator. In fact, in the first case (IaaS), the operator will offer a cloud environment, where the industry vertical may install any kind of edge platform (e.g., in collaboration with suppliers and system integrators); instead, with the PaaS model, the operator is already offering an MEC environment, where the vertical may directly deploy applications (again, possibly in cooperation with suppliers and system integrators). Instead, the third model is essentially adopted in the extreme case where the vertical is using directly a complete software solution offered by the operator (i.e., including both MEC platform and MEC applications). For that reason, we have analyzed more in detail the main cost structures related to the two models, considered as most probable ones (PaaS and IaaS), as follows:

- **PaaS** – with this model, both 5G network infrastructure and MEC servers (including an MEC platform) are hosted by MNO and offered to the industry vertical. In this case, the vertical may neglect CAPEX components of the overall costs, and sustain only OPEX, which usually range from design to operation and maintenance of SW instances (e.g., MEC applications, and related management and orchestration).
- **IaaS** – with this model, both 5G network infrastructure and MEC servers (excluding an MEC platform) are hosted by MNO, and offered to the industry vertical. In this case, the vertical may need to install and operate not only MEC applications, but also an MEC platform in the target MEC hosts of the MEC system. This also implies the need to manage and orchestrate the whole MEC system. Here, many subcases are possible. One of them is the following: since this IaaS model requires a more complex cost management, the vertical may decide to leverage on a collaboration with suppliers and system integrators, who may take the responsibility of the MEC system layer, while the vertical may concentrate its efforts on the application layer only.

8.4 Case Study: The Automotive Sector

A complete and detailed automotive case study is very complex and is subjected to many assumptions, since that industry sector is composed of many players, with different strategies and regional specificities (see, e.g., the numerous 5GAA membership). Nonetheless, an exemplary analysis of the main aspects related to MEC deployments, from an automotive perspective, could be useful as a starting point, also to better understand the MEC potential in this domain. In the following, we start from a brief description of V2X technologies, to discuss general aspects of 5G and benefits given by MEC.

8.4.1 V2X Technology Landscape

When we talk about V2X, the attention immediately focuses on the recent debate around access technology adopted to enable V2X communications. In particular, short-range direct communication between cars (called also sidelink communication) can be realized through two competing standards existing in the field: WAVE/DSRC based on the IEEE 802.11p standard and the LTE-based 3GPP feature LTE C-V2X. The latter also enables long-range communication using traditional cellular networks operating over the mobile network operator licensed spectrum. Standard bodies have independently specified these two technologies (DSRC and C-V2X), which rely on different channel access schemes. Consequently, the main debate is about regulation and the usage of ITS band (e.g., license-free spectrum in the 5.9 GHz band), where the two technologies are supposed to access. Here, in fact, the role of regulation is to guarantee that multiple technologies can coexist in the ITS band, by defining the rules for handling coexistence, interoperability, and backward compatibility issues. In Europe, the current orientation for the 5.9 GHz ITS band follows the principle of technology neutrality, but the debate is still open, as the regulations are still under finalization. In the United States, a similar discussion is ongoing with the objective to enable DSRC as well as LTE C-V2X to access the 5.9 GHz band. Apart from a neutral usage of the spectrum, the debate is also around performances (and related communication range offered by the two competing technologies) and other performance metrics, other than cost aspects for road operators and mobile operators. The debate is quite complex and conducted by many stakeholders (thus not only regulators, but also car makers, technology providers, etc.), even if there is quite a consolidate number of industry players more in favor of C-V2X, as a more performing and future-proof technology. As an example, again, the huge

5GAA membership is a clear sign that the majority of the automotive industry is oriented for the usage of 5G cellular infrastructure. In any case, at this point in time, it is hard to predict how the debate will be closed. In any case, it is reasonable to imagine that, for global adoption of V2X services, the market will need to converge toward an interoperable solution, supporting multi-MNO, multi-OEM, and multi-vendor scenarios.

In this complex context, the role of MEC (which is in principle an access-agnostic technology) is quite orthogonal and independent from these debates. In fact, in principle, the MEC architecture is valid for any kind of access technology (DSRC and C-V2X). Nevertheless, it should be also noted that edge servers (which are expected to be deployed in the operator's infrastructure) need to be connected to UEs (cars) through a cellular network interface. So, it is also quite clear how C-V2X is in general a better solution for MEC deployment.

8.4.2 Benefits of MEC for 5G Automotive Services

Talking about 5G, it is quite clear how telecommunication companies see automotive use cases as strictly related to the usage of C-V2X, since this technology exploits their infrastructure and gives them new business opportunities. Also, the majority of car makers are oriented toward cellular-based connectivity, and committed to influence 5G standardization in 3GPP, in order to meet the urgent needs and performance requirements coming from automotive services. Indeed, this is also the goal of associations like 5GAA, where a large community of players (putting together the two worlds of telecommunication and automotive sectors) [109] is building consensus to bring relevant technical requirements and suitable solutions in 3GPP. On the other hand, ETSI MEC, representing the only international standard available on edge computing topic, is working on the different verticals enabled by MEC, and also working on the specification of an MEC V2X API [110], as suitable V2X service, aiming at facilitating the V2X interoperability in a multi-vendor, multi-network, and multi-access environment.

Figure 8.5 shows an example of such a scenario. In particular:

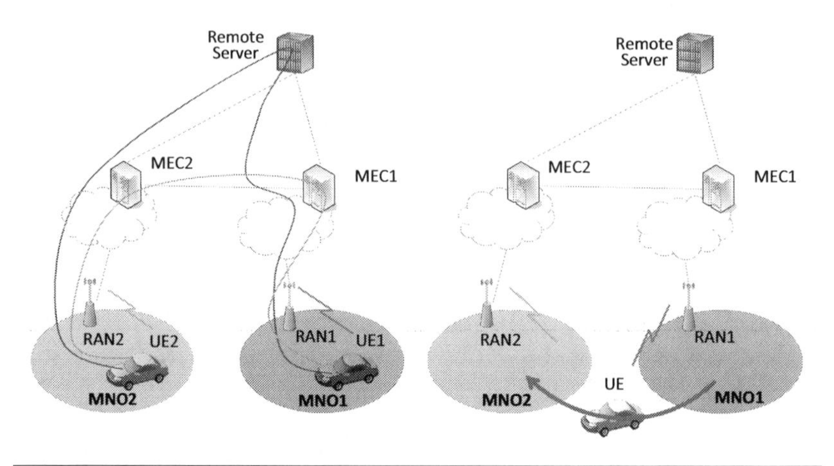

Figure 8.5 Example of MEC deployment for V2X in multi-operator environments [111].

- On the left-hand side: in red, the data path is considering the interconnection between MNOs terminated at the remote side, and is causing high end-to-end (E2E) latency. In green, instead, a direct interconnection between MNOs is envisaged in order to reduce significantly the E2E latency. This means that a standardized "interface" (e.g., MEC V2X API) would support communication between vehicle and application in edge cloud, and also between vehicles belonging to different OEMs.
- Right-hand side: the car is temporarily out of coverage of two operators, and there is the need to provide V2X service continuity across all the territory including both the areas (i.e., also moving from Operator#1 to Operator#2), without service disruption and by ensuring E2E connection. MEC can be a solution to host side-link configuration and manage a multi-operator environment, especially when the vehicle is out of coverage. So, again a standardized "interface" (e.g., MEC V2X API) would help in order to correctly manage side-link configuration parameters, when the vehicle is out of coverage.

In summary, it is clear how MEC is in general a preferred technology for automotive stakeholders, not only with benefits in terms of end-to-end performance in V2X services, but also as a solution to enable interoperability among different players in the ecosystem.

8.4.3 MEC Use Cases for 5G Automotive Services

In a recent white paper, the 5GAA categorized a comprehensive list of connected vehicle applications, by giving some first examples of MEC-relevant use cases [65], under the following categories:

- Safety (e.g., intersection movement assist (IMA)),
- Convenience (software updates),
- Advanced driving assistance (e.g., real-time situational awareness & HD maps),
- Vulnerable road user (VRU).

An example of safety use cases is the IMA, where the driver is warned of collision risk through an intersection. In this case, MEC systems could provide support for real-time data analysis, data fusion, and reduced ingress bandwidth with respect to the remote cloud.

In other use cases, like real-time situational awareness & HD maps, the driver is alerted of hazard (icy) road conditions, on the basis of information collected from the surrounding entities (e.g., other cars, sensors, roadside infrastructure, network). The aim is to improve traffic flow, traffic signal timing, routing, variable speed limits, weather alerts, etc. These use cases collect the most challenging requirements for V2X, from an MEC perspective: here, the MEC system can be requested to host the collection, processing, and distribution of a relatively large amount of data with high reliability and low latency in parallel.

MEC is also an ideal solution for use cases like detection of VRU, especially in the view of exploiting local context and information available at the edge. In particular, published APIs like radio network information (RNI) and location APIs can help in improving the accuracy of the positioning information of all traffic participants.

Notes

1 When talking about Internet of Things, some people also refer to Fog.
2 A careful reader may notice that sometimes, a company can be, in principle, categorized in more than one vertical. In addition to that, it may happen that companies expand their business and start focusing on areas where initially they were not present. For these reasons, the above categorization should not be considered as a limitation, but just as a tool to identify the main impacts on MEC coming from very heterogeneous requirements.

3 On the other hand, data transport saving is also one important advantage of edge computing. Nevertheless, this aspect should be considered more as a benefit from the point of view of the operator, which is in fact operating the communicating network, instead of the industry vertical.

PART 3

MEC Deployment and the Ecosystem

9
MEC Deployments
Cost Aspects

The present chapter includes a very detailed cost analysis for multi-access edge computing (MEC), considering different possible edge cloud business models, and also from different points of view. An edge-specific cost structure is also described for a few cases of interest, with particular focus on the operator point of view. Finally, some MEC business model aspects are discussed.

9.1 Infrastructure Evolution

As we have seen, the introduction of edge computing is a change of paradigm for many stakeholders, with respect to current business models. In particular, in terms of actual market adoption, the decision on MEC deployment depends on the infrastructure owner (operator, vertical industry, cloud provider, etc.). In order to better understand the business opportunities, it is thus important to analyze costs associated to the different deployment options, depending on the ownership and related business models. In fact, decision-making is always strictly related not only to revenue opportunities but also to the sustainability of the associated costs. In this chapter, we do not aim at providing a complete business plan, but at least aim to analyze cost aspects related to edge computing deployment. The starting point will be a description of the current trend in the evolution of data centers, communication network, and devices. This will enable a better understanding of edge cloud deployments and cost models. Then, an exemplary total cost of ownership (TCO) analysis will be followed by the introduction of the business model canvas, as a suitable tool to provide a complete overview of the aspects related to the MEC market adoption.

9.1.1 Data Center Evolution

In an era dominated by global data traffic increase, the evolution of data centers plays a key role, even with the adoption of edge computing technology. In fact, it is important to first clarify that the introduction of edge servers is not going to replace huge data centers, as this new paradigm is not in contrast with centralized management of data traffic in huge data centers, deployed remotely. Instead, MEC is a complementary technology, helping to expand the footprint of cloud computing toward the edge of the network, and also enabling evolutionary concepts like distributed computing [88].

In fact, although the portion of traffic residing within the data centers still represents a great majority (about 75%) of the global IP traffic [89], we have to consider that according to the forecasts, traffic between data centers is expected to grow until 2021 (and even faster than either traffic to end users or traffic within the data center). This will lead to a couple of key aspects:

- on the one hand, data centers will continue to evolve, to satisfy the increasing traffic demand, and move toward better performance and lower costs;[1]
- on the other hand, a more connected world is needed in the near future, with better efficiency in the communication between clouds (enabling distributed computing).

As a consequence, an understanding of the evolution of data centers is a key starting point for the definition of future edge cloud solutions. In particular, a recent and consolidated trend for the evolution of data centers is represented by Hyperscale Computing, which is defined as follows (https://en.wikipedia.org/wiki/Hyperscale_computing):

In computing, hyperscale is the ability of an architecture to scale appropriately as increased demand is added to the system. [...]. This typically involves the ability to seamlessly provision and add compute, memory, networking, and storage resources to a given node or set of nodes that make up a larger computing, distributed computing, or grid computing environment. Hyperscale computing is necessary in order to build a robust and scalable cloud, big data, map reduce, or distributed storage system and is often associated with the infrastructure required to run large distributed sites such as Facebook, Google,

Microsoft, Amazon, or Oracle. Companies like Ericsson, Advanced Micro Devices and Intel provide hyperscale infrastructure kits for IT service providers.[2].

Thus, hyperscale data center (or computing) refers to facilities and provisioning required in distributed computing environments. From a technology view, this evolution is driven by the increasing requirements for application performance and by the growing needs to reduce costs (operational and capital expenditures, i.e., OPEX and CAPEX), as well as ever-increasing technology investments. Nevertheless, it should also be kept in mind that many companies are engaged in the evolution toward more efficient data centers, and implementing the cloud as a partial or complete solution that could help them rapidly scale their infrastructure. This is a huge and heterogeneous set of companies, ranging from enterprises and service providers to governments (thus, not just "hyperscale" companies).

More in detail, the main architectural shifts in the evolution of data centers, impacting the key building blocks of network/switching, compute/server, and storage are briefly listed in the following.

- Networking speed is a critical performance requirement; while a consolidated level is around 8 Gbps, the industry is looking at better interconnections (from 16 to 32 Gbps).
- Storage technology and protocols are evolving, and flash is becoming more prevalent to drive high-density packaging and performance.
- Thermal constraints and flexibility/scalability needs are also driving rack architecture changes. A recent introduction of the Rack Scale Design (RSD) technology[3] is offering server disaggregation and modular utilization of resources, providing benefits in terms of flexibility, cost reduction per unit of capacity.
- Usage of advanced cooling technologies (e.g., free cooling) and increase of room temperature (enabled by new HPC equipment) are providing huge savings in terms of energy consumption.
- Silicon architectures and integrated circuits: many cloud companies are utilizing different silicon architectures such as graphics processing unit (GPUs), for example, to accelerate the computing speed for single instruction multiple data (SIMD) processing. More recently, tensor processing units

(TPU) have also emerged, to run specific tasks, for example, for the elaboration of machine learning (ML) algorithms.

- Also, acceleration through field programmable gate array (FPGA) is a technology used in many servers for security and performance reasons (in particular, for highly computational applications, such as financial trading, specialized voice and data analytics). The advantage of FPGA also resides on programmability, since the array of gates that make up an FPGA can be programmed on demand (after they are manufactured, delivered, and deployed), to run a specific algorithm, in order to meet very specific workload requirements.

The last two aspects mentioned in this list are particularly relevant for edge computing, since computational power for edge task processing becomes even more critical at the edge. In fact, if we assume to deploy an MEC server in proximity to end users (thus, with very low end-to-end packet transfer delay), the proportional impact of processing delay is increased, especially for latency-critical applications (e.g., robotics, industrial, VR/AR but also big data and elaboration from huge number of Internet of Things [IoT] devices). Thus, especially for these use cases, packet processing evolution is highly important to support the introduction of edge computing.

An example of a high-demanding task driving the evolution of processing units is given by ML algorithms and more in general by artificial intelligence (AI). For this purpose, it's worth mentioning Google's recent announcement of the introduction of Edge TPU (a small-footprint ASIC chip designed to run TensorFlow Lite ML models on edge devices) and Cloud IoT Edge (the software stack that extends Google's cloud services to IoT gateways and edge devices).[4] Moreover, these tasks can also be accelerated through FPGA, providing high efficiency in terms of power consumption, which is an important enabler for the adoption of edge computing (thus, in small-footprint cloud servers).

9.1.2 Communication Network Evolution

Moving the point of view from IT cloud providers to mobile operators, it is important to describe how communication network infrastructure

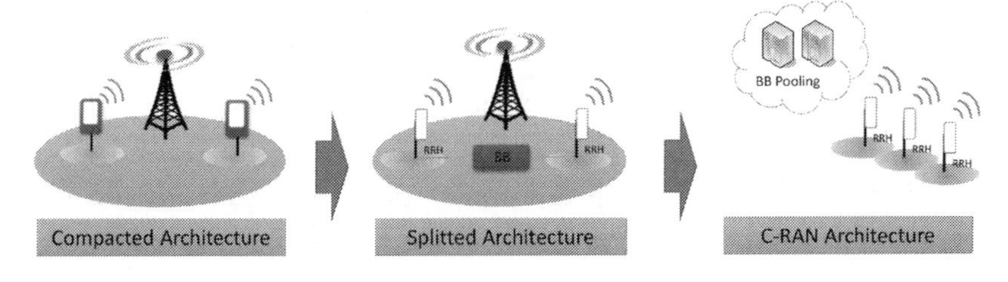

Figure 9.1 Mobile network evolution path toward C-RAN.

is evolving, in order to better understand the context in which edge computing is expected to be deployed.

In particular, we can identify the main steps of this evolution path [91], summarized also in Figure 9.1:

- Traditionally, mobile networks started from the so-called **compacted architecture**, consisting of the baseband (BB) unit and the radio unit joint in the same module;
- The subsequent step was thus the adoption of a **splitted architecture** (with BB unit and radio unit in different modules), that can be seen as a first centralization step toward a Cloud-RAN (C-RAN) architecture;
- Recently, many mobile operators started to adopt **C-RAN architecture**, where BB resources shared (pooled) across RRUs.

The main driver for this infrastructure evolution is cost saving, where energy consumption and OPEX are an important cost item, considered by mobile operators with increasing attention (especially considering the whole telecommunication network). In more detail:

- Literature studies have already provided methods [100] to evaluate the different consumption drivers in a mobile site (mainly considering radio base station and transmission part). Nevertheless, as we well know, in practical cases, operator costs are also influenced by infrastructure within mobile sites, and for this reason, the introduction of 5G networks is highly influenced by the introduction of Cloud-RAN (C-RAN) as a new architectural paradigm able to save energy, thanks to the adoption of a greener centralized infrastructure (see also [101]).

- Some more recent studies [103] have also assessed the overall consumption by considering not only the RAN part, but also the infrastructure equipment present in mobile sites (air conditioning, power supply, and so on), showing that centralizing BB processing in C-RAN environments may lead to significant savings, with respect to a typical flat architecture [102,104].

After these preliminary steps toward C-RAN architecture, the current trend followed by operators in the framework of 5G is now the migration toward a **Virtual RAN** (V-RAN). In this phase, virtualization (on general-purpose hardware [HW]) enables the abstraction from a particular operating system. This is usually done by a "hypervisor." Some functionalities are executed running as virtual machines (VMs), and the virtualization approach preferably follows the general ETSI network function virtualization (NFV) framework. Figure 9.2 shows an example of V-RAN implementation, where multiple VMs represent single radio access techniques (RATs), for example, 2G and 3G, or by subsystems of the protocol stack of a single RAT (e.g., PHY, MAC, RLC).

Talking about the advantages of Virtual RAN for the operator, we can say that V-RAN has all the C-RAN advantages, in addition to the following (due to usage of general-purpose hardware):

- the operator can dynamically allocate processing resources within a centralized BB pool to different virtualized base stations and different air interface standards
- HW and SW are totally decoupled, in terms of both cost and management

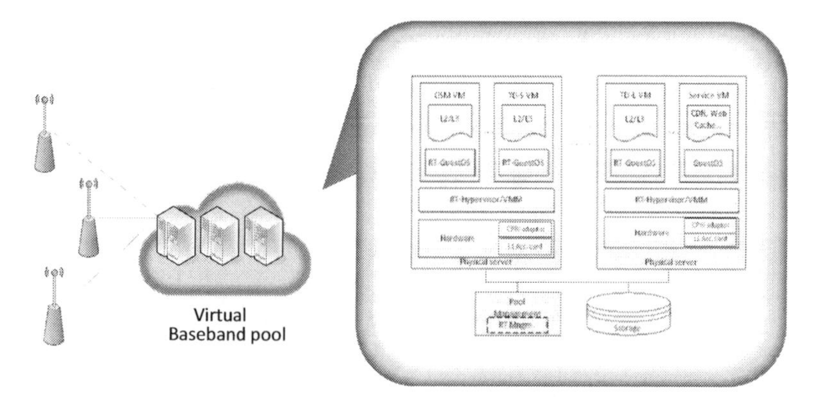

Figure 9.2 Example of Virtual RAN implementation.

- simpler inter-vendor interoperability
- cost reduction to manage, maintain, expand, and upgrade the base station.

Many challenges continue to be associated with V-RAN implementation: in general, to satisfy real-time needs given by radio systems, it may be difficult to implement all eNBs protocol stack on a general-purpose hardware (HW); in these cases, some functionalities (typically running in the lower part of the PHY layer) need to be implemented on dedicated HW, for example, FPGA-based acceleration cards. In any case, full RAN centralization implies the usage of fiber (with high capacity, but also most expensive); so, if this cable infrastructure is not available, partial RAN centralization solutions are usually evaluated by operators. In these cases, flexible solutions of partial RAN split are preferred, that is, where the processing partitioning between RRH and BB elements is flexible, and can be changed overtime (e.g., thanks to FPGA programmability).

9.1.3 Devices Evolution

When it comes to mobile devices, it is clear in everybody's mind how their evolution in these decades saw a huge technology gap, between their very first introduction as voice terminals to current smartphones [92,93]. This evolution in fact covers many aspects, including dramatic improvements from the point of view of computational capacity, storage, display, user interface, and of course data connectivity (both cellular, and also related to other radio accesses, e.g., Wi-Fi, Bluetooth, and NFC).

More in particular, when it comes to computational power, these smartphones are currently equipped with high-performance applications, like voice assistants (e.g., Apple's Siri, Google's Assistant, Microsoft's Cortana, Amazon's Alexa), real-time translation between languages, or again high-quality photo/video shooting, with real-time and automatic optical processing.

Some of these complex tasks are made possible by the introduction of technologies of AI and ML, which are already used by the present-day smartphones. In addition to that, according to some forecasts,[5] in 2020, more than a third of all smartphones on the market will have built-in AI capabilities.

Another example of high-demanding applications is given by AR/VR, which together with ML technologies are widely recognized as the main pillars of the digitalization of industries [94]. In typical AR applications, the user is usually equipped with eye-tracking smart glasses [95] and tactile gloves rather than screwdriver sets [96].

Running all these high-demanding tasks would result in high battery consumption and heat dissipation. At the same time, end-to-end latency requirements (e.g., few milliseconds) do not allow the running of the complete application in large central data centers, due to the physical limits of light speed in optical fibers. For these reasons, it could result convenient deploying some application components at the network edge, in order to offload the device while maintaining short latency. This is the well-known task-offloading use case [97].

In summary, it is clear how edge computing is a key supporting technology, even in an era with evolved and powerful terminals. This aspect should not surprise the reader, because essentially the evolution of terminals complexity cannot cover alone all the challenging needs and stringent requirements of very complex tasks and applications foreseen in the coming years.

Another important direction for the evolution of terminals is due to the progressive introduction in the market of IoT services, and the related presence of a huge number of heterogeneous terminals, sensors, wearable and IoT devices.

Nowadays, devices like Amazon Echo, Google Chromecast, and Apple TV are powered by content and intelligence that resides in the cloud. These exemplary devices are already entering the customers' homes, and are even expected to have an increasing role in the future, for example, acting as gateways toward IoT sensors and a myriad of small- and low-end devices. Here, edge computing will offer again important benefits in terms of local processing of big data coming from smart devices, providing also better speed and higher bandwidth with respect to larger data transfers to remote data centers.

9.2 Edge Cloud Deployment Options

In this infrastructure evolution scenario, the described distributed computing scenario offers to telecom operators many degrees of freedom for the deployment of edge clouds. As depicted in Figure 9.3, with the

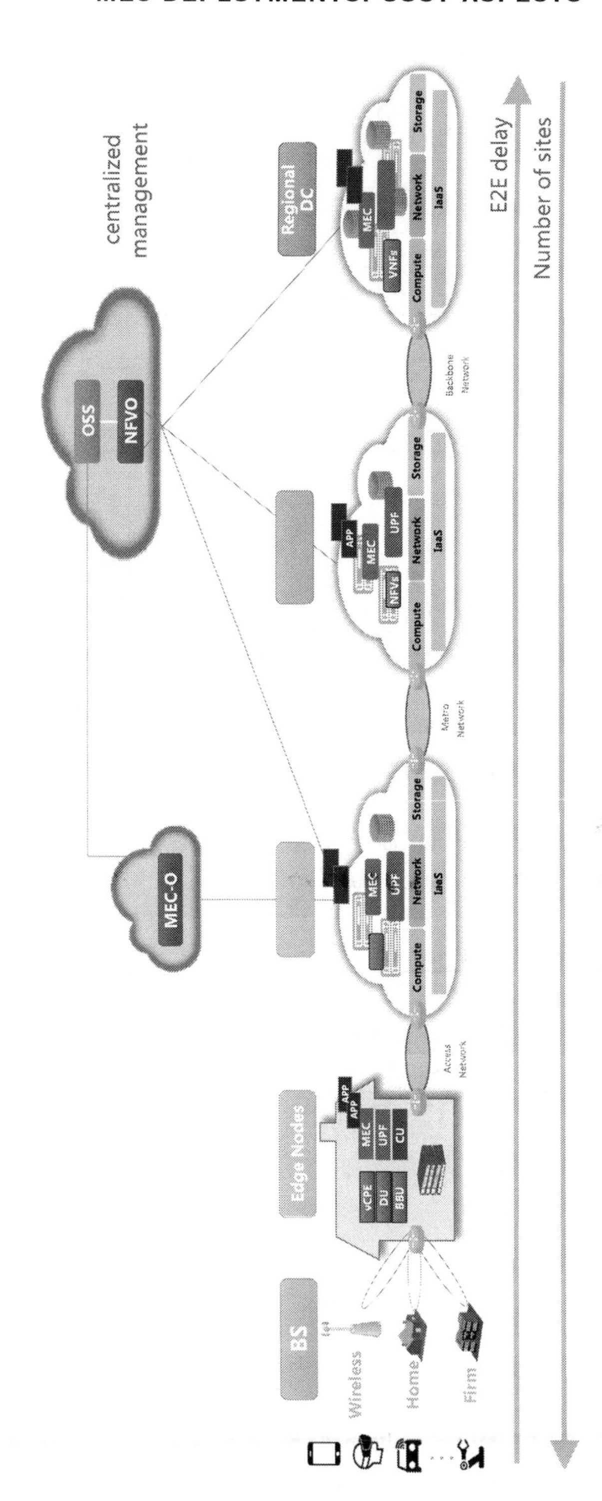

Figure 9.3 Different edge cloud levels.

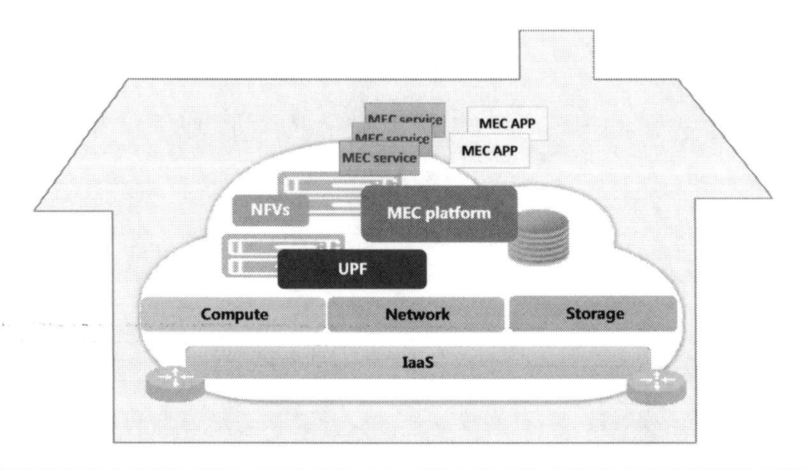

Figure 9.4 Edge data center (general case).

lower level of clouds, the offered E2E delay is lower (left-hand side of the picture), but at a prize of a bigger number of sites; instead, a higher E2E delay is offered by a small number of bigger DCs (right-hand side).

This fundamental trade-off has already been described earlier in this book, and the actual choice among different deployment options essentially depends on the infrastructure owner's decision on the edge cloud business model. All main models are described in the following section, together with their associated cost structure, which has an impact on business convenience, both in terms of CAPEX and OPEX of the overall infrastructure (depicted in Figure 9.4). Nevertheless, it is worth noting that sometimes the decision of the operator can be also driven by other motivations, for example, due to liability, security, or again strategic ownership and control, to mitigate the risk of disintermediation from the other stakeholders.

9.3 Edge Cloud Business Models

In the following subsections, many offering models (traditionally present in literature for cloud computing) are listed and analyzed, with particular reference to edge cloud specificities.

9.3.1 Real Estate Model

The simplest offering model (called Real Estate [RE] model here) is essentially the rental of all edge facilities and gives mostly all

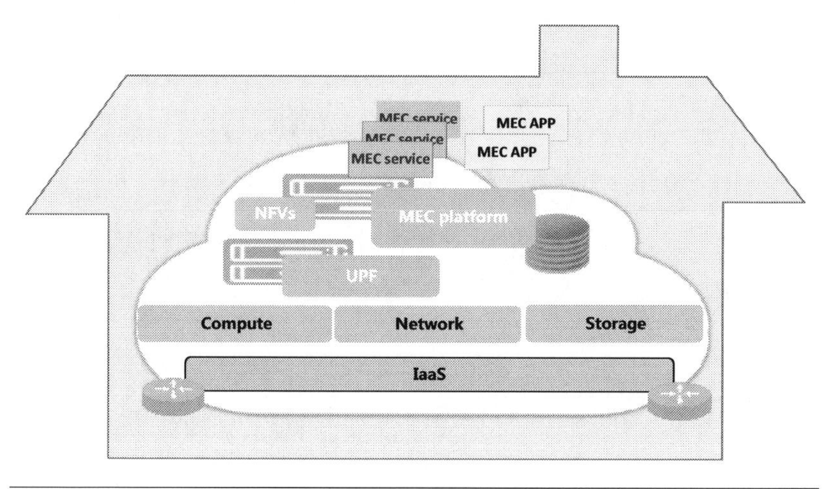

Figure 9.5 Edge data center: example of RE model.

the freedom to the MEC infrastructure provider. This may include (or not) room cooling, utilities (e.g., with uninterruptible power supply (UPS)), and external connectivity in the data center (both wired and eventually wireless, e.g., through radio links).

This model is associated with a trivial cost structure, with no specific aspects related to edge computing, except from the fact that network connection at the edge site should meet certain requirements in terms of availability, speed, and latency, otherwise the tenant will never deploy the MEC infrastructure (because of potential issues with the end customers on MEC application performances) (Figure 9.5).

It is not clear to which extent this model is the preferred one by mobile operators, rather than instead by tower companies (e.g., providing basic facilities and generic infrastructure for many operators and technology providers). In any case, with this simple model, all the responsibility of the final MEC service performance is left to the tenant, and this could be an advantage in case the owner is not interested in taking any risks.

9.3.2 Infrastructure-as-a-Service (IaaS)

With this model, both the physical and virtual infrastructure are provided to third parties, thus offering not only room facilities, cooling, and external connectivity in the data center, but also virtual hardware resources (compute network and storage) that could be

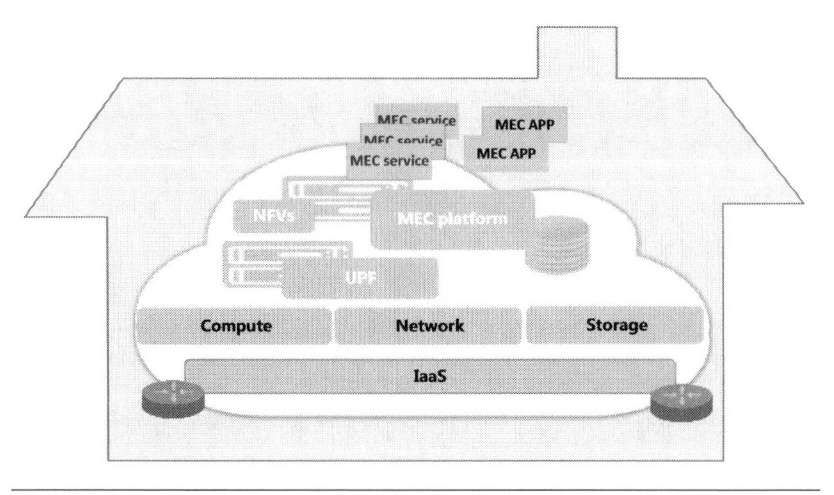

Figure 9.6 Edge data center: example of IaaS model.

exploited not only by MEC providers, but in general by any cloud provider (Figure 9.6).

As we can see easily, there isn't anything specific to the edge cloud in this model, except the fact that some hardware infrastructure may contain specific functionalities or characteristics (e.g., FPGA acceleration) that are essential for running edge services. Also, in the case of Infrastructure-as-a-Service (IaaS), the decision related to the actual MEC deployment is subjected to the specific deal conditions between the infrastructure owner and the tenant (which may need to impose, e.g., to prevent potential issues with the end customers on MEC application performances). On the other hand, with this simple IaaS model, all the responsibility of the final MEC service performance is left to the tenant, and this could be an advantage in case the owner is not interested in taking big risks.

9.3.3 Platform-as-a-Service (PaaS)

With this model, not only virtual resources (compute network and storage) are provided to third parties, but also an MEC platform and eventually other network functionalities (depicted below as VFNs). In fact, depending on the owner (which can be an operator or simply a cloud provider), the network functionalities may be (or not) present in the edge data center, or in any case may be (or not) exposed to the tenant (Figure 9.7).

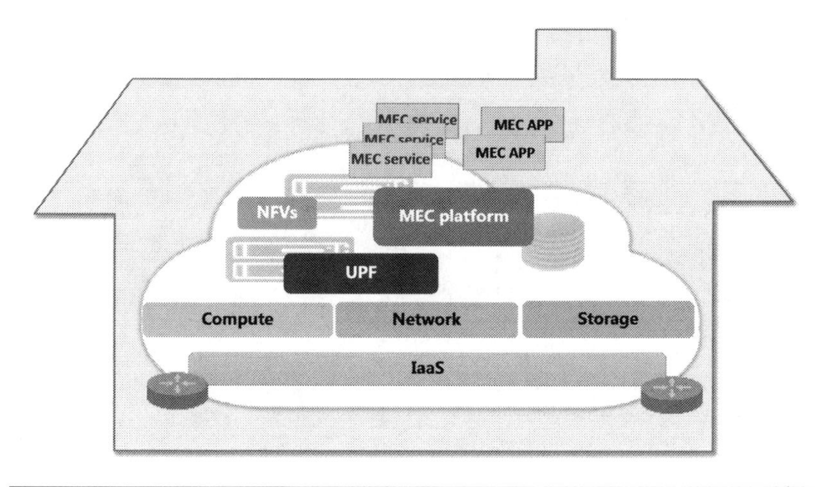

Figure 9.7 Edge data center: example of PaaS model.

This model is also particularly suitable for hosting multiple tenants, as the virtual infrastructure offered is quite complete, and includes many functionalities that can be exploited by different stakeholders. This also comes with more responsibility from the owner side, which will have to take care of the costs associated to management, orchestration, operation, and maintenance of all the offered infrastructure.

9.3.4 Collaborative Platform-as-a-Service (C-PaaS)

This model is a variant of the PaaS, where the owner may be interested in offering as well some value-added services, for example, through collaboration with third parties and companies specialized in different fields (like financial services, big data analysis, or content distribution network (CDN) platform and content providers, system integrators for automotive or industrial automation, etc.). In this case, the collaborative approach with verticals and other industrial players provides a rich set of functionalities to application developers, for example, with a set of (RESTful) APIs organized in a sort of middleware exposed at the application level (Figure 9.8).

This is the most open and collaborative model, associated of course with the highest level of costs, which include (in addition to the items given by previous business models) also the management of the MEC platform to ensure full operational add-value MEC services. In addition, the owner takes the responsibility of also guaranteeing the actual performance of offered services. On the other hand, more

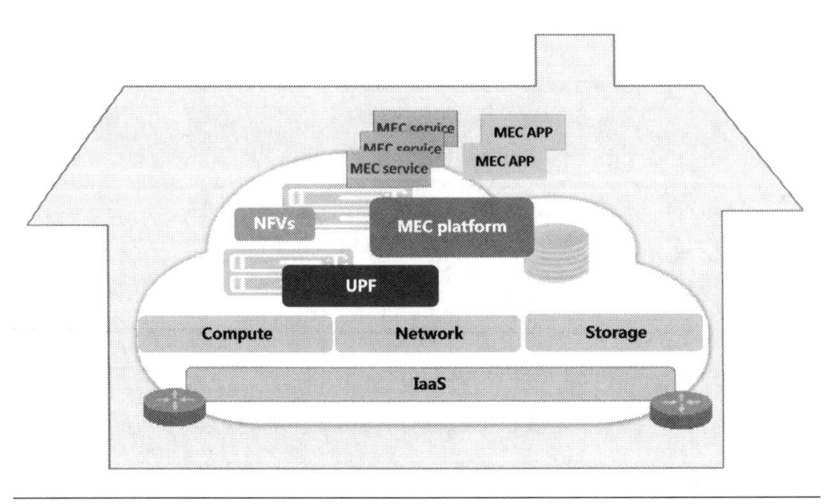

Figure 9.8 Edge data center: example of C-PaaS model.

revenue opportunities are also associated with this model, which is characterized by the highest level of control and the richest portfolio of functionalities offered to customers. As a last note, similarly to Software-as-a-Service (SaaS), this model permits third parties to build their own services and get exposed to higher level application developers, paving the way for the creation of a richer edge developers' ecosystem.

9.4 TCO Analysis (PaaS Model): An Operator Perspective

In the following we will analyze the cost structure (from an operator point of view) related to the PaaS model. In fact, among the previously described edge cloud business models, this is the most straightforward case of collaboration between MNO and MEC application developers, and at the same time with the relatively simple cost structure (excluding the trivial case of a "real estate" model). To analyze the PaaS, we should start from the IaaS case, which essentially considers the cost items related to owning an edge data center infrastructure. Evaluating the data center cost model is not straightforward, as a complete analysis would include many aspects.

In general, there are six cost categories [98]: network, head count, servers, facilities, OS and management, and storage and backup and recovery (showed in Figure 9.9).

Recent studies evolved this financial model from a project- and component-based model to a more holistic unit-costing model [98],

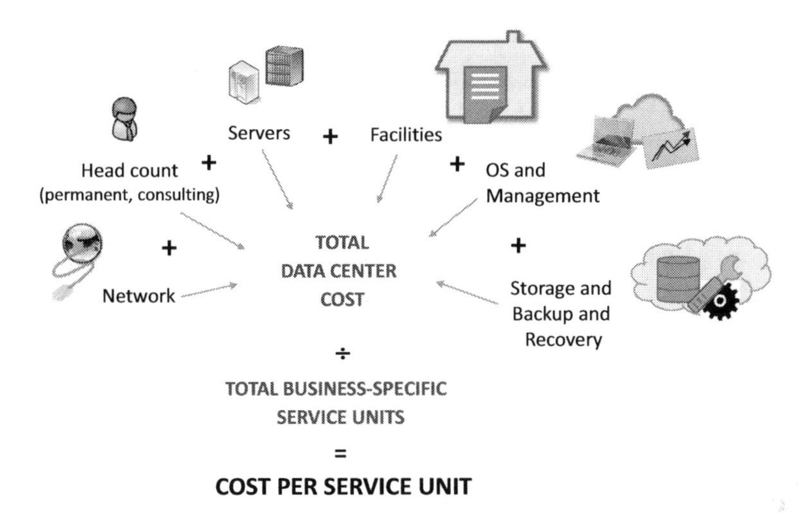

Figure 9.9 Example of calculation of normalized cost per service unit. (Source: Intel [98].)

where, for example, a cost per service unit is derived (here, all categories of cost are considered and divided by the number of units for that environment, such as EDA-MIPS of performance per system). The advantage of this approach is to have a more effective benchmark of data center performance and cost assessment, for example, to prioritize data center investments.

The PaaS model adds to IaaS essentially all costs specifically related to the MEC platform. In this case, there may be several variants, as the operator may decide to deploy MEC in a stand-alone NFVI infrastructure, or to use the same general-purpose hardware (which is supposed to be already present for NFV deployments) and to offer in addition the MEC components under the same NFVI. This second case, from a cost perspective, appears to be the most convenient, even if the reader should consider that operators may take very different decisions, depending on country-specific situations, or management costs related to specific deal conditions, or again other business reasons (e.g., liability, security). For this reason, the exact cost evaluation will depend on the specific deployment and operator choices. Nonetheless, Table 9.1 provides quite a comprehensive list of the main cost items for the PaaS model for edge cloud offering, divided per category (CAPEX and OPEX).

In particular, the costs (both CAPEX and OPEX) associated with the physical and virtual infrastructure are very relevant for the operator, and for that reason, the decision to deploy a new edge site needs

Table 9.1 Edge Cloud Cost Structure (PaaS Model)

CAPEX (CAPITAL EXPENDITURES)	
COST ITEM	COMMENTS
Edge site facilities	See RE model
NFVI, VIM	See IaaS model
MEC platform	Purchasing of basic MEC platform (ETSI GS MEC 011), optionally with selected MEC APIs (ETSI GS MEC 009)
MEPM, MEC-O, OSS	Integration of MEC platform in the network and OSS/BSS
OPEX (OPERATIONAL EXPENDITURES)	
COST ITEM	COMMENTS
Cost of operation staff, consultants	Including management, operation, and maintenance of virtual infrastructure
Sales and marketing costs	Specific edge offers depend on the customized functionalities and targeted customers
License fees of MEC server	This can be provided by a third party, expert in the field
Building/footprint of the server	Maintenance and costs related to physical infrastructure
Energy costs	Electricity (e.g., cooling) is a very relevant cost item

necessarily to take into account the actual convenience and revenue opportunities (see Section 9.5 for the business model canvas). For that reason, MEC is often considered as co-located with NVF sites, and as a part of a bigger infrastructural renewal plan, that is, related to the virtualization of the whole operator's network, toward the introduction of 5G systems. In this perspective, the decision about MEC deployment is just involving those cost elements related to MEC platform and services, implemented as VFNs under the common virtualization infrastructure (NFVI) and management and orchestration (MANO).

In order to have a more clear idea of the actual savings in the MEC business plan, the next subsection provides a short overview of the energy costs, for example, mainly associated with room cooling.

9.4.1 Focus on Energy Costs: Cooling Aspects

Energy consumption in a mobile network is a relevant part of the operational costs of an operator, especially due to the huge number of sites needed to cover the entire country [99]. In addition to that, China Mobile showed in an analysis of the TCO of mobile sites [102] that OPEX accounts for over 60% of the overall TCO (while the CAPEX only accounts for about 40% of the TCO), and that electricity and site rent represent the great majority of operational expenditures (respectively, 41% and 32% of the OPEX).

To give an idea of the order of magnitude of the site TCO (for the more general case of a data center), a world-class performance is often considered to be less, 0.10 USD/kW/h total cost of ownership (including all facilities, staffing, energy, equipment depreciation, etc.), while more advanced cooling solutions (sometimes also customized for edge computing[6]) may enable deployments with facility/capex costs of under 0.02 USD/h. In that perspective, energy consumption is a key factor for the decision on (edge) cloud deployment.

9.4.2 Deployment Costs: MEC in NFV Environments

Since the steps associated with the progressive network virtualization (i.e., through NFV) are part of a huge plan for operators, involving costs associated with the overall infrastructure renewal, the decision to adopt the NFV framework, is often considered a mandatory step for them, in order to keep operational efficiency and maintain their competitiveness in the 5G era.

On the other hand, the decision to deploy MEC is also driven by other factors (e.g., new revenue streams generation), and aligning the roadmap for the introduction of MEC to the bigger NFV path (by mainly considering MEC co-located with NFV sites) could be a convenient and cost-effective way to upgrade the network. Table 9.2 thus summarizes the main cost items related to the introduction of MEC in NFV environments (opposed to the general case, already described in Table 9.2, of MEC–stand-alone deployments).

As we can see, since MEC components are essentially implemented as VFNs under the common virtualization infrastructure (NFVI) and management and network orchestration (MANO), the cost structure is significantly simpler, as related to the acquisition, integration, and management of these components (namely, MEC platform and MEC service APIs).

Finally, as a side (but important) remark, since MEC deployment can be mitigated by the co-location with NFV, when it comes to operators, the decision to deploy MEC components could be less constrained by cost factors and more driven by actual business and service

Table 9.2 Edge Cloud Cost Structure (PaaS Model) – MEC in NFV Case

CAPEX (CAPITAL EXPENDITURES)	
COST ITEM	COMMENTS
MEC platform	Purchasing of basic MEC platform (ETSI GS MEC 011), optionally with selected MEC APIs (ETSI GS MEC 009)
MEPM, MEC-O, OSS	Integration of MEC platform in the network and OSS/BSS
OPEX (OPERATIONAL EXPENDITURES)	
COST ITEM	COMMENTS
Cost of operation staff, consultants	Including management, operation, and maintenance of virtual infrastructure
Sales and marketing costs	Specific edge offers depend on the customized functionalities and targeted customers
License fees of MEC server	This can be provided by a third party, expert in the field

needs. Incidentally, the creation of new services was also the natural reason for the creation of MEC.

9.5 Business Model Aspects (Operator Perspective)

As we have shown earlier in this book, edge computing is expected to unlock 25% of the total 5G revenue potential for operators. On the other hand, we have already clarified that the MEC ecosystem will be more articulated than a simple mobile business, as operators will not be the only stakeholders that may capture value from MEC (thus, other players such as verticals, over-the-top (OTT) players, and developers will most likely join the party). For this reason, it is not easy to evaluate the total addressable market (TAM), and only preliminary assessments have been done so far. As an example, analysts [87] have provided some initial forecast of U.S. and Western Europe installations for MEC, revealing that the largest number of MEC appliances (in an earlier future) are expected to be in the retail industry, followed by other sectors like health care, manufacturing, transportation, and warehousing (while MEC installations are expected to come later). This should not surprise the reader, since the study is simply calculating the number of potential MEC locations as directly correlated to the retailer facilities (e.g., a national retailer may have 2,000 locations each with a small MEC appliance), while one single manufacturing plant (which may employ the same number of people as in 2,000 retail locations), could be supported with a single, large MEC installation.

9.5.1 Exemplary Business Model Canvas

As we've seen, depending on the business model, the MEC cost structure may vary significantly. Moreover, the revenue potential of MEC is in general not limited by MEC installation, but could also include value-added services offered by edge cloud application developers and third-party content providers in the 5G era. For these reasons, it is not straightforward and immediate to conduct a serious and honest business study. For the sake of accuracy, a specific evaluation of business opportunity is left to the reader. What instead we provide here is a business model canvas,[7] as a simple (but effective) tool often used to evaluate all aspects related to a new business.

Figure 9.10 shows the template of a Business Canvas Model, which is nothing but a representation of how an organization makes (or intends to make) money. This model is articulated in nine building blocks (from value proposition to client segments, communication/distribution channels, client relationships, key resources, activities, partners, revenue streams, and costs).

Finally, when it comes to MEC business, this model should be customized, having in mind the specific deployment option, the target stakeholders, and (most importantly), the specific service enabled by MEC. In fact, as we have seen from multiple angles in this book, at the end is the service which influences MEC performance, by imposing

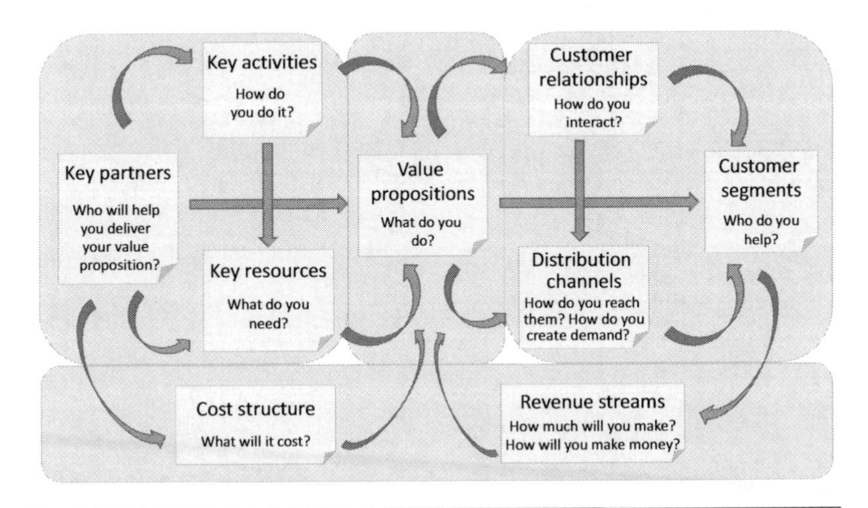

Figure 9.10 Template of Business Canvas Model. (Source: Wikipedia.)

end-to-end requirements to network and edge cloud infrastructure, and also determines deployment options, with consequent impact on cost structure and opportunity to introduce MEC in the system.

Notes

1 For further deepening, Ref. [90] provides a detailed analysis of the improvements and cost savings enabled by this relatively recent data center strategy over the years.
2 NOTE: this movement is relatively recent; however, the reader should keep in mind that hyperscale computing is not just an American phenomenon; it is global, with leading players around the world (e.g., including companies well-known under the acronym of BAT, i.e., Baidu, Alibaba, and Tencent, driving significant innovation across Asia).
3 www.intel.com/content/www/us/en/architecture-and-technology/rack-scale-design-overview.html
4 More info for Cloud IoT Edge can be found here: https://cloud.google.com/iot-edge/
5 https://root-nation.com/gadgets-en/smartphones-en/en-ai-in-smartphones/
6 www.cloudcooler.co.uk/edge-computing
7 https://en.wikipedia.org/wiki/Business_Model_Canvas

10

THE MEC ECOSYSTEM

In this chapter, we provide a complete overview of the huge ecosystem of edge stakeholders, categorizing them based on their actual usage of MEC technology.

The chapter ends with a description of the ETSI ISG MEC activities for ecosystem engagement, including proof of concepts (PoCs), trials, and Hackathons, together with a brief description of open source projects and research communities.

10.1 MEC: A Heterogeneous Ecosystem of Stakeholders

In the second part of this book (Chapters 5–8), we have analyzed the MEC market from different perspectives. More widely, the MEC ecosystem is composed of a bigger set of stakeholders, and heterogeneous type of players, which have a role in the adoption of MEC technologies. Figure 10.1 depicts the main sectors composing the huge MEC ecosystem.

As we can see, this ecosystem is quite diverse and complex, and on the other hand, it is worth identifying some high-level categories, based on their actual usage of MEC technology. For this purpose, we can identify the following main profiles of stakeholders in the MEC ecosystem:

- Operators and Service Providers
- Traditional Vendors (providing Telco infrastructure)
- IT Infrastructure Providers (generic IT Cloud)
- Verticals and System Integrators (including SMEs/start-ups)
- Software Developers (including not only over-the-top (OTT) and big companies but also start-ups, SW houses, application developers, research communities, etc.)

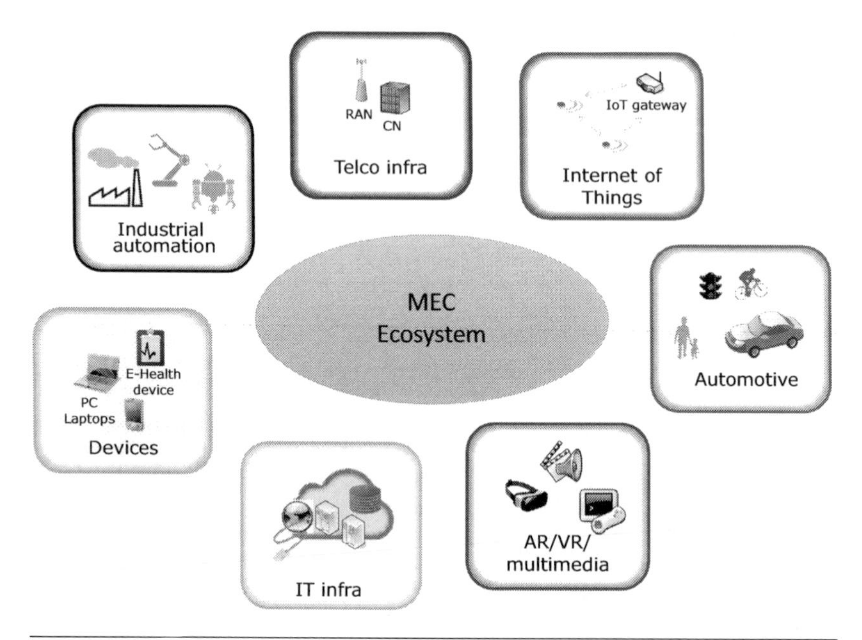

Figure 10.1 Sectors of the MEC ecosystem.

Of course, these profiles should be seen just as generic categories of stakeholders, as in principle, for each of them we can have individuals or companies coming from different sectors of the MEC ecosystem. Nonetheless, this classification (used in the present chapter) is useful to better understand the specific use of MEC that a specific stakeholder is doing, and what is the impact on the development of the MEC ecosystem. In the end, what matters is the market, which is composed of different players (not all necessarily involved from a commercial perspective, e.g., open source developers).

In summary, the success of MEC depends on how market leaders and decision makers will be able to break the famous and well-known chicken-and-egg problem:

- Typically, operators and infrastructure owners could complain about the lack of applications and ecosystem of developers, and use this as a motivation to delay their MEC deployment plans and the consequent offering of their infrastructure.

- On the other hand, application developers could complain about the lack of MEC infrastructure (where instantiating, testing, and running their apps), as a motivation to not implement new application and services at the edge.

This typical chicken-and-egg problem is already well-known by most of the companies, as the market is starting to realize that a collaborative approach is needed for the success of MEC. Companies that traditionally were in competition are now working together (e.g., in standardization organizations or industry groups, or open source projects) in order to provide solutions and means to boost the adoption of edge computing technologies. This collaborative approach, which is well-known since many years but was actually adopted only recently by companies, is also known as *open innovation*,[1] and is recognized as key to the creation of new services in complex environments such as IT and telecommunications.

10.1.1 Operators and Service Providers

This category of stakeholders is identified as one of the first decision makers able to break the chicken-and-egg problem, as they are most likely the first (or the main) infrastructure owners interested in introducing edge services in their networks. Of course, they are not the end customers (which are instead represented not only by vertical industry players, like automotive and industrial automation, but also by public administration, road operators, consumers, etc.). On the other hand, operators and service providers are expected to have a role in the MEC adoption.

Moreover, mobile operators and service providers are already progressively defining their strategy for the adoption of edge computing, and their positioning and motivation [54] for the adoption of edge computing is becoming more and more specifically defined and elaborated, with respect to past years.

Figure 10.2 shows two main branches as main business drivers for MEC (from an operator perspective): **revenue generation** and **cost saving**. The first one (revenue generation) was in fact the original motivation for the creation of ETSI ISG MEC (mobile edge computing) [55, 56] as a standardization group different from ETSI

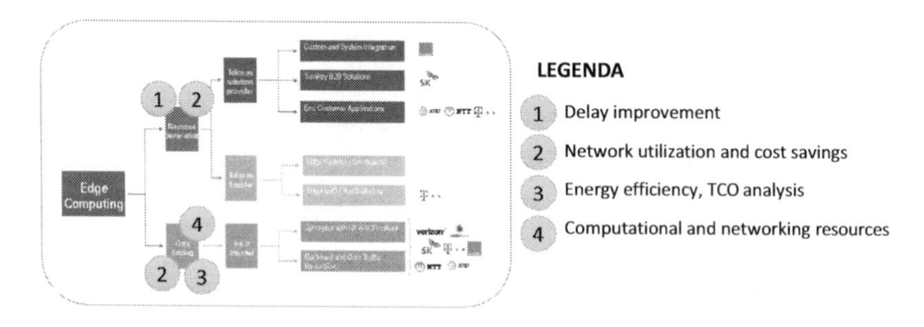

Figure 10.2 KPIs for MEC and impact on business. (Elaboration from Ref. [54])

ISG NFV (network function virtualization) [57], which was instead mainly motivated by the second driver (cost saving).

In fact, on the one hand, *cost saving* was (and still is) the main driver for the full virtualization of network infrastructure. Decoupling (general purpose) hardware from software permits the mobile operators and infrastructure owners to efficiently manage and operate the NFV-based network, save costs for upgrades and maintenance, have a more flexible control of the deployment, and also avoid silos-oriented implementations.

On the other hand, edge computing (being besides preferably based on a virtual infrastructure as well) was certainly seen by operators as a means for providing them other additional benefits, in terms of *revenue generation*, and the consequent possibility to enable new and added-value services (e.g., thanks to low latency, real-time radio access network and context information locally and at the edge of the network). These services could be provided by Telco as a solution provider or as a middleware for third parties, through suitable new business models, for example, based on Platform-as-a-Service (PaaS) offerings to application developers and OTT stakeholders.

In summary, we can certainly say that the landscape of the different MEC drivers (derived from the various business motivations for its adoption) is currently quite diversified; in particular, the following aspects can be highlighted as main KPIs for infrastructure owners, mobile operators, and service providers:

- **Delay improvement**, in terms of better end-to-end (E2E) quality of experience (QoE) perceived by the end users, thanks to proximity;

- **Network utilization and cost savings** (due to, e.g., more efficient usage of backhauling and transport network, as the edge server is conveniently routing locally use traffic);
- **Energy Efficiency and total cost of ownership** (TCO), thanks to the small footprint of edge servers, and consequent better improvements in site costs management (especially when room cooling can be avoided, or outdoor deployments can be foreseen);
- **Management of computation and networking resources,** as essentially MEC software instances are conveniently deployed and are run on top of virtualized infrastructure, possibly hosting also other VNFs and processing tasks.

As a consequence, it is quite clear that KPIs for MEC are quite diversified, and they include of course not only the latency (even if this one is the most commonly recognized KPI by the community of bloggers, journalists and mass communication).

The usage of MEC by operators is influenced by these KPIs. In fact, they tend to start from the knowledge of the technology to primarily understand how to deploy MEC in their NFV environment. Then, they usually assess the different solutions in PoCs or trials, to better see the actual benefits of MEC in their labs, or in the field.

At the same time, some operators are also taking more "active" roles, and evaluating the development of MEC application programming interfaces (APIs) and middleware services, to improve the quality of their network, or to offer services to third parties and application providers.

In all these cases, operators have clearly recognized the importance of standardized solutions (e.g., through RESTful APIs) that will enable the portability of MEC solutions across different platforms and help application developers to interoperate even between different MEC systems.

In some other cases, operators such as DT are even creating spin-off companies (e.g., MobiledgeX), with the explicit purpose of developing edge computing solutions as autonomous business goals. These companies are also involved in ecosystem engagement activities, such as MEC Hackathons,[2] as suitable tools to attract developer communities and encourage them adopting MEC.

10.1.2 Telco Infrastructure Providers

Typically, the ecosystem of infrastructure providers comes from a traditional situation where there were complete end-to-end solutions offered to operators, fully implemented in dedicated hardware. More recently, the progressive virtualization of network functions obliged them to change their way of working, by providing network functionalities as VFNs capable of running on general-purpose hardware. With the advent of ETSI ISG MEC (in 2014), it was clear how some technology providers started to abandon that traditional mindset. Today, since edge computing is already part of the bigger 5G picture, we can say that many network infrastructure vendors are considering service exposure as a tool to enable new services not only for their traditional customers (operators) but also for third parties or even application developers.

In all these cases, most of the companies are also involved in open source projects, for example, for the development of management and orchestration frameworks, with the goal to facilitate the integration work in charge to the operator, or again in other communities for the promotion of their MEC APIs and functionalities.

Also, in this category of stakeholders, some infrastructure providers sometimes consider the creation of separate companies for the engagement of application developers.[3]

10.1.3 IT Infrastructure Providers

Usually, IT providers come from consolidated cloud businesses in huge (and remote) data centers, already having realized that the future of the cloud resides on the edge. Examples of these companies are Intel, HPE, IBM, Juniper Networks, Fujitsu, etc.

Expanding their footprint to the edge cloud is definitely a new business opportunity for these IT companies, as more servers can be potentially sold, thanks to MEC deployments. Nevertheless, one of the barriers is of course given by the lack of applications and SW reference frameworks or suitable software development kits (SDK) that may trigger this market. For that reason, some of these companies are engaged in open source communities and projects aiming at developing the software needed to create the edge cloud infrastructure. Examples of these communities are given in Sections 10.4 and 10.5.

10.1.4 Verticals and System Integrators

This category of stakeholders includes those companies working in specific vertical industry domains, which are identified by operators as one of the main drivers for 5G. Thus, we can say that verticals are actually B2B customers for operators and represent the main reference for the identification of key use cases relevant to MEC.

In this category, we have also inserted system integrators and Tier-1 suppliers, since these players work in close collaboration with verticals for the practical realization of technical solutions that are going to be implemented into products. As an example for the automotive sector, usually the car makers (e.g., Ford, BMW, Daimler, Audi) work with system integrators and Tier-1 suppliers (e.g., Denso, Bosch, Continental); thus, when it comes to actual usage of MEC technology to develop edge services, it is reasonable to consider both these groups of companies. And by the way, most of the time, involving them in experimental activities or PoCs/trials is key for the success of those initiatives.

This sector (taking again the automotive example) was also historically characterized by a static attitude in terms of willingness to change production systems or adopt new IT technologies. Nevertheless, with the advent of autonomous and connected cars, these companies started to change mindsets, and become more resolutely engaged in collaborations and industry groups (e.g., 5GAA, AECC). As a proof of change of perspective, some companies also abandoned definitely the concept of car ownership, starting to collaborate with other car makers for offering more advanced evolved services. A typical example, related to car sharing, is given by BMW (initially providing DriveNow) and Daimler (with Car2go) that started to collaborate together to offer a joint service in Germany (called Share Now[4]).

In other cases, Tier-1 suppliers are fully engaged in small consortia, together with car makers, operators, and Telco infrastructure providers, to drive innovative solutions using edge computing. This is the case of a recent MEC trial, where Continental, Deutsche Telekom, Fraunhofer ESK, MHP, and Nokia successfully concluded some tests of connected driving technology on the A9 Digital Test Track.[5] The activity, based on the Car2MEC project (funded by Bavarian Ministry for Economic Affairs, started 2016), proved that MEC is critical for driving safety on the path to fully automated driving, by showing that time critical information can be delivered from one car to another in less than 30 ms in an LTE network combined with an MEC-based edge cloud.

10.1.5 Software Developers

Starting from the actual definition of MEC, it is quite clear how software developers are a key stakeholder for this technology and for its success in the market:

Multi-access edge computing offers to application developers and content providers cloud-computing capabilities and an IT service environment at the edge of the network.

In addition, by recalling the previously described chicken-and-egg problem, we could even argue that software developers are a key "customer" of MEC technology. In fact, a great portion of the success and the adoption of MEC is given by the actual creation of an ecosystem of applications developers. In fact, we could imagine a wide adoption of MEC only if a consistent deployment of MEC servers is supported by a good number of applications, running at the edge of the network and providing innovative and value-added services.

The first barrier for the engagement of developers is the absence of a development kit, or a software reference platform that could help them to develop their apps, and guide them for the usage of MEC technologies. A first key example of software frameworks made available for developers is given by the Internet of Things (IoT), where Google offers **Cloud IoT Edge** as a tool for developing applications running on devices at the edge of the network, for example, to manage IoT sensors, analyze data, and communicate with a central cloud (see Figure 10.3).

Other companies are also expanding their IoT domain to the edge. In fact, at their re:Invent developer conference[6] in 2016, Amazon announced the usage of their AWS Greengrass and Lambda (serverless computing) offering, to extend Amazon Web Services (AWS) to intermittently connected edge devices.[7] The introduction of

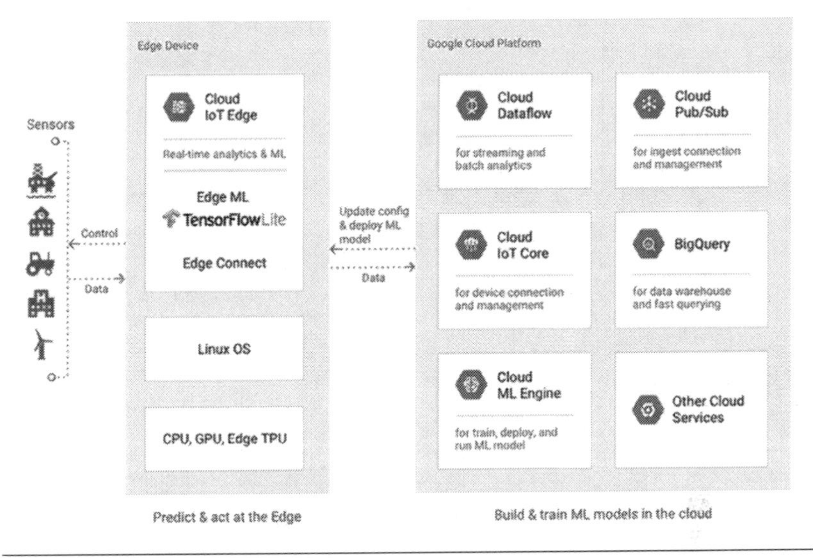

Figure 10.3 Cloud IoT edge. (Source: https://cloud.google.com/iot-edge/.)

Lambda@Edge permits developers to run their code globally at AWS locations close to users so that they can deliver full-featured, customized content with high performance and low latency.

> *"With AWS Greengrass, developers can add AWS Lambda functions to a connected device right from the AWS Management Console, and the device executes the code locally so that devices can respond to events and take actions in near real-time. AWS Greengrass also includes AWS IoT messaging and synching capabilities so devices can send messages to other devices without connecting back to the cloud," said Amazon. "AWS Greengrass allows customers the flexibility to have devices rely on the cloud when it makes sense, perform tasks on their own when it makes sense, and talk to each other when it makes sense – all in a single, seamless environment."*

Another example of environments offered to software developers in the IoT space is given by **Azure IoT Edge**, which allows cloud workloads to be containerized and run locally on smart devices ranging from a Raspberry Pi to an industrial gateway. This was introduced at Microsoft's BUILD 2017 developer conference and was announced to be generally available since June 2018.

As depicted in Figure 10.4, Azure IoT Edge comprises three components: IoT Edge modules, the IoT Edge runtime, and IoT Hub.

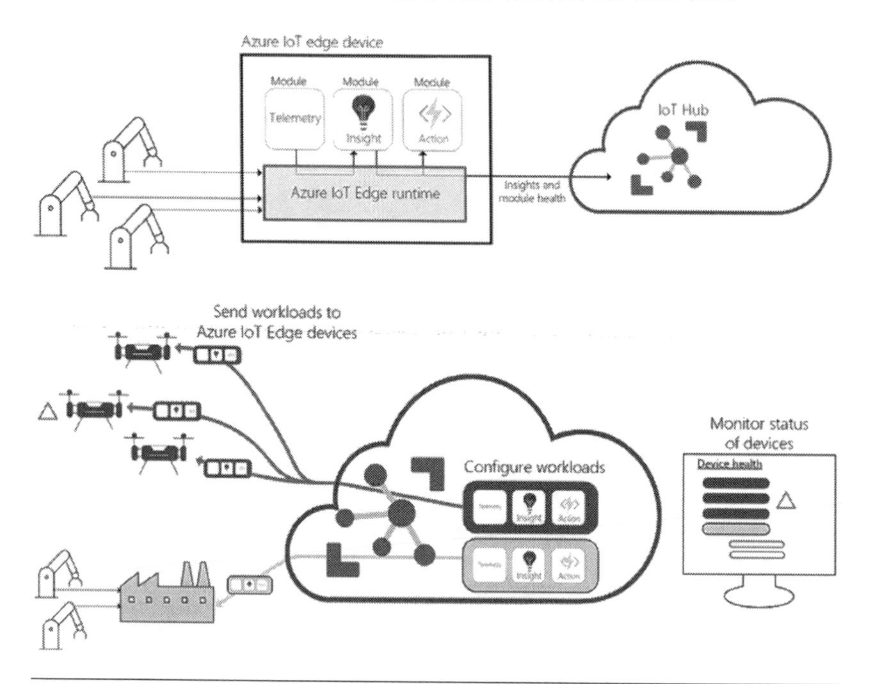

Figure 10.4 Azure IoT edge. (Source: Microsoft.)

IoT Edge modules are containers that run Azure services, third-party services, or custom code; these modules are deployed to IoT Edge devices and execute locally on them. The IoT Edge runtime runs on each IoT Edge device, managing the deployed modules, while the IoT Hub is a cloud-based interface for remotely monitoring and managing IoT Edge devices.

10.2 Ecosystem Engagement in ETSI ISG MEC

As a general assumption, an important role for the engagement of the ecosystem of applications developers is given by the presence of standardized solutions. Of course, standards are traditionally not the best place to engage SW developers, essentially because in the past years they were engaged more in open source communities, while standard bodies introduced functionalities outside these environments, and worked at a different speed and flexibility, often not enough for the needs of software developers.

Nevertheless, the ETSI ISG MEC is one of the virtuous exceptions in this depicted scenario, as the huge ecosystem of companies involved

Figure 10.5 The ecosystem of ETSI ISG MEC members and participants.

in that group (Figure 10.5) recognized the need to place side by side the traditional standards activities on MEC with ecosystem engagement activities. In the following section, we describe few exemplary ETSI ISG MEC activities going into this direction of better engagement of a diverse set of stakeholders and to bridge the gap with developers.

10.2.1 MEC Proof of Concepts

The first initiative agreed in the ISG was the creation of an MEC PoC framework [105], where the group defined a methodology[8] for submitting, approving, and managing PoC in ETSI ISG MEC. Proofs of concept are in fact an important tool to demonstrate the viability of a new technology and provide feedback to the standardization work.

The MEC PoCs (listed also in the MEC wiki page: https://mecwiki.etsi.org) are intended as multiparty projects showcasing early implementations of few selected MEC components, the results of which are fed back to the ISG, for a proper feedback on the normative work. Nevertheless, these are not intended as MEC-compliant implementations: in fact, neither ETSI, its ISG MEC, nor their members make any endorsement of any product or implementation claiming to demonstrate or conform to MEC (and no verification or test has been performed by ETSI on any part of these MEC PoCs).

10.2.2 MEC Hackathons

The second step in the ecosystem engagement was done by the ISG MEC, with the approval of an MEC Hackathon Framework (also described in the MEC wiki[9]), for a proper definition, evaluation, approval, and management of MEC Hackathons.

According to the general definition, a hackathon is a design event in which application developers (including graphic, UX, interface, protocol designers, and project management) collaborate in order to develop a new service or application within a limited amount of time. When it comes to MEC, Hackathons are events organized with the goal to promote knowledge and adoption of ETSI MEC ISG standards, and in particular, to encourage all stakeholders to use MEC work to develop edge applications, especially by utilizing MEC service APIs.

10.2.3 MEC Deployment Trials

The last framework created by the ISG was related to the MEC Deployment Trials (MDTs).[10] These are evolutions of PoCs, where MEC technologies are demonstrated in commercial trials/deployments, thus not anymore as simple lab/prototype implementations.

The list of active MDTs is again published in the MEC wiki, and for each approved MDT, the team is expected to demonstrate their MDT proposal at a public event, for example, Public Exhibition, ISG MEC meeting, or other events, by means outlined in their own MDT proposal.

10.2.4 MEC DECODE Working Group

More recently, the ETSI Multi-access Edge Computing group (ETSI MEC ISG) announced the creation of the Deployment and Ecosystem Development working group (WG DECODE).[11] This WG will focus on accelerating the market adoption and implementation of systems using the MEC-defined framework and services exposed using MEC-standardized APIs. Among the activities of DECODE, we can mention the following:

- Facilitating the use of open source components for the implementation and validation of MEC-related use cases or MEC system entities.
- Identifying best practices to implement an MEC System, by leveraging cloud application design, orchestration and automation, security and reliability advances.

- Enabling operator adoption and interoperability by developing and maintaining specifications related to testing, including guidelines and API conformance specifications.
- Increasing the accessibility and adoption of MEC specifications by exposing OpenAPI (aka Swagger)-compliant MEC API descriptions via the ETSI Forge site, and in doing so, opening up the ecosystem to third-party application developers.

10.3 Industry Groups

Vertical associations are key stakeholders in the MEC ecosystem. As we have seen previously in this book, many industry groups (e.g., 5GAA, AECC, 5G-ACIA, VR-IF, VRARA) are working on the different vertical market segments and consider MEC as a key supporting technology for their respective use cases.

Nevertheless, in most of the cases, these industry groups so far didn't make extensive use of MEC technology, but still are at an early stage of use cases elaboration and definition of technical solutions. On the other hand, the reader should also be aware that the market is still in an early stage of 5G deployment, and thus should not be surprised about the related state of maturity of MEC adoption by industry groups. Overall, the expectation is that verticals will adopt MEC in conjunction with the progressive introduction of 5G, at least because any credible performance evaluation (conducted from their point of view) should be preferably based on real implementations, and thus will need to collaborate with 5G network operators.

10.4 Open Source Projects

A totally different story is traced by open source communities, which are traditionally playing a key role in the design of software frameworks and for the related creation of an ecosystem of applications. This phenomenon is also true for edge computing, where MEC apps can start to appear only when an MEC reference implementation is released as an open source and provides satisfying guidelines for SW developers. Few examples are already active so far, and in the following subsection, we provide a suitable reference for developers.

10.4.1 Akraino Edge Stack

This is a Linux Foundation project (initiated by AT&T and Intel in 2018) that aims to create an open source software stack that supports high-availability cloud services optimized for edge computing systems and applications [106]. In principle, Akraino scope covers everything about edge, that is, from the development of edge solution to address Telco, Enterprise, and Industrial IoT, to the development of Edge API and framework for interoperability with third-party edge providers & hybrid cloud models, to the development of Edge Middleware, SDKs, applications, and create an app/VNF ecosystem. Due to its nature, Akraino Edge Stack contains several integration projects (or Blueprints) and feature projects. A blueprint is a concept used by Akraino to define a declarative configuration of the entire stack, that is, edge platform that can support edge workloads and edge APIs. In practice, in order to address specific use cases, a reference architecture is developed by the community, by specifying all the components used within that reference architecture (e.g., HW, SW, tools to manage the entire stack, and point of delivery, i.e., the method used to deploy in a site).

Among the feature projects, we can mention the **MEC API Framework**,[12] which is essentially a collection of mechanisms to enable applications in a distributed cloud, offering offers services that bring applications and services together by allowing to applications offering or consuming services either locally or remotely. In fact, applications hosted in the distributed cloud, that is, the edge and central cloud, can consume services offered by service producers. Service consumers can discover the services that are available in that location via the API framework. Similarly, the service producers can advertise their offerings via the same API framework. In addition to service discovery, the API framework allows authentication and authorization and can also provide communications transport to the service consumers and producers.

10.4.1 OpenStack Foundation Edge Computing Group

This is a new group created by OpenStack Foundation [107], with the aim to drive the evolution of OpenStack to support cloud edge computing. Based on initial community interest expressed at the OpenStack Summit Boston (a two-day workshop held in September 2017), the OSF Edge Computing Group has identified several challenges and has started to work on fundamental requirements of a fully functional edge computing cloud infrastructure, also in collaboration with other edge-related projects.[13]

According to the identified challenges, edge resource management systems should deliver a set of high-level mechanisms whose assembly results in a system capable of operating and using a geo-distributed Infrastructure-as-a-Service (IaaS) infrastructure relying on WAN interconnects. In other words, the challenge is to revise (and extend when needed) IaaS core services to deal with aforementioned edge specifics – network disconnections/bandwidth, limited capacities in terms of compute and storage, unmanned deployments, and so forth.

More in detail, more work is needed in the following directions:

- A virtual-machine/container/bare-metal manager in charge of managing the machine/container life cycle (configuration, scheduling, deployment, suspend/resume, and shutdown).
- An image manager in charge of template files (aka virtual-machine/container images).
- A network manager in charge of providing connectivity to the infrastructure: virtual networks and external access for users.
- A storage manager, providing storage services to edge applications.
- Administrative tools, providing user interfaces to operate and use the dispersed infrastructure.

10.5 Research Communities

The role of collaborative projects and research communities is key to the development of solutions for edge computing. In particular, there are many EU-funded projects under Horizon2020 framework which are driving the innovation toward 5G systems, as they are composed by heterogeneous consortia (i.e., including not only outstanding universities and research centers, but also SMEs and big industrial players, such as operators and technology providers). These 5G research projects are also coordinated by the 5G Infrastructure

Public Private Partnership (5G PPP), which is a joint initiative between the European Commission and the European ICT industry (ICT manufacturers, telecommunications operators, service providers, SMEs, and researcher institutions) with the goal to deliver solutions, architectures, technologies, and standards for the ubiquitous next-generation communication infrastructures of the coming decade [108]. Few examples of recent 5G-PPP projects are:

- 5G-Media (www.5gmedia.eu/), focused on integrated programmable service platforms for the development, design, and operations of media applications in 5G networks, and using ETSI MANO framework;
- 5GCity (www.5gcity.eu/), working on the integration of both ETSI MEC and ETSI NFV architectures and interfaces, with the aim to design, implement, and deploy a distributed cloud, edge, and radio platform for smart cities and infrastructure owners acting as 5G Neutral Hosts;
- 5G Essence (www.5g-essence-h2020.eu/) addressing the paradigms of Edge Cloud computing and Small Cell as a Service through an edge cloud environment based on a two-tier architecture (the first distributed tier for providing low-latency services and the second centralized tier for providing high processing power for computing intensive network applications);
- Matilda (www.matilda-5g.eu/), where multi-site management of the cloud/edge computing and IoT resources is supported by a multi-site virtualized infrastructure manager;
- 5G-Coral (http://5g-coral.eu/), leveraging on the pervasiveness of edge and fog computing in the radio access network (RAN) to create a unique opportunity for access convergence;
- 5G-Transformer (http://5g-transformer.eu), aiming at transforming today's rigid mobile transport networks into an SDN/NFV-based mobile transport and computing platform, by also envisioning an edge computing platform capable of offering services tailored to the specific needs of vertical industries.

In addition to EU collaborative projects, there are also international trial and testing activities which are often developing MEC technologies as key components of experimental activities. In fact, often these international players are also involved in initiatives, for example, with the aim to create and open research and innovation laboratories focusing on 5G systems, where in most of the cases, edge computing is a critical technology to support the use cases of interest. This is the case of 5Tonic (www.5tonic.org/), a laboratory based in Madrid, created also with the goal to promote joint project development and entrepreneurial ventures, discussion fora, events, and conference sites in an international environment.

Notes

1 https://en.wikipedia.org/wiki/Open_innovation
2 https://tmt.knect365.com/edge-computing-congress/etsi-mec-hackathon
3 https://edgegravity.ericsson.com/application-providers/
4 www.your-now.com/our-solutions/share-now
5 www.continental-corporation.com/en/press/press-releases/2019-03-21-car2mec-168160
6 https://aws.amazon.com/lambda/edge/
7 These should be intended as "smart" edge devices, of course. In fact, Greengrass requires at least 1 GHz of compute (either Arm or x86), 128 MB of RAM, plus additional resources for OS, message throughput, and AWS Lambda execution. According to Amazon, "Greengrass Core can run on devices that range from a Raspberry Pi to a server-level appliance".
8 https://mecwiki.etsi.org/index.php?title=PoC_Framework
9 https://mecwiki.etsi.org/index.php?title=MEC_Hackathon_Framework
10 https://mecwiki.etsi.org/index.php?title=MEC_Deployment_Framework
11 www.etsi.org/newsroom/press-releases/1548-2019-02-etsi-multi-access-edge-computing-opens-new-working-group-for-mec-deployment
12 https://wiki.akraino.org/display/AK/MEC+API++Framework
13 https://wiki.openstack.org/wiki/Edge_Computing_Group

References

1. E. Dalhman, et al., *4G, LTE-Advanced-Pro and the Road to 5G*, 3rd Ed., Academic Press, Cambridge, MA, 2016.
2. M. Olsson, et al., *EPC and 4G Evolved Packet Networks*, 2nd Ed., Academic Press, Cambridge, MA, 2013.
3. J. Cartmell, et al., Local Selected IP Traffic Offload Reducing Traffic Congestion within the Mobile Core Network, in *Proceedings IEEE CCNC 2013*, Las Vegas, Nevada, USA, 2013.
4. SCF046, "Small Cell Services." Available at: http://scf.io/en/documents/046_Small_cell_services.php.
5. SCF084, "Small Cell Zone Services: RESTfule Bindings." Available at: http://scf.io/en/documents/084_-_Small_Cell_Zone_services_RESTful_Bindings.php.
6. SCF091, "Small Cell Application Programmer's Guide." Available at: http://scf.io/en/documents/091_-_Small_cell_application_programmers_guide.php.
7. SCF014, "Edge Computing Made Simple." Available at: http://scf.io/en/documents/014_-_Edge_Computing_made_simple.php.
8. NIST Special Publication 500-325, "Fog Computing Conceptual Model: Recommendations of the National Institute of Standards and Technology," 03/2018. Available at: https://doi.org/10.6028/NIST.SP.500–325.
9. NGMN, "5G White Paper," 2015. Available at: www.ngmn.org/fileadmin/ngmn/content/downloads/Technical/2015/NGMN_5G_White_Paper_V1_0.pdf.
10. ETSI, "Mobile Edge Computing: A Key Technology Towards 5G," 2015. Available at: www.etsi.org/images/files/ETSIWhitePapers/etsi_wp11_mec_a_key_technology_towards_5g.pdf.

11. ETSI, "MEC Deployments in 4G and Evolution Towards 5G," 2018. Available at: www.etsi.org/images/files/ETSIWhitePapers/etsi_wp24_MEC_deployment_in_4G_5G_FINAL.pdf.
12. ETSI, "Developing Software for Multi-Access Edge Computing," 2017. Available at: www.etsi.org/images/files/ETSIWhitePapers/etsi_wp20_MEC_SoftwareDevelopment_FINAL.pdf.
13. Nassim Nicholas Taleb, *Antifragile*, Random House, New Yok, 2014.
14. M. Sifalakis, et al., An Information Centric Network for Computing the Distribution of Computations, in *Proceedings ACM-ICN 2014*, Paris, France, 2014.
15. ETSI GS MEC 003, "Mobile Edge Computing (MEC); Framework and Reference Architecture," v. 1.1.1, 03/2016. Available at: www.etsi.org/deliver/etsi_gs/MEC/001_099/003/01.01.01_60/gs_MEC003v010101p.pdf.
16. ETSI GS MEC 011, "Mobile Edge Computing (MEC); Mobile Edge Platform Application Enablement," v. 1.1.1, 07/2017. Available at: www.etsi.org/deliver/etsi_gs/MEC/001_099/011/01.01.01_60/gs_MEC011v010101p.pdf.
17. ETSI GS MEC 009, "Mobile Edge Computing (MEC); General principles for Mobile Edge Service APIs," v. 1.1.1, 07/2017, Available at: www.etsi.org/deliver/etsi_gs/MEC/001_099/009/01.01.01_60/gs_MEC009v010101p.pdf.
18. ETSI GS MEC 012, "Mobile Edge Computing (MEC); Radio Network Information API," v. 1.1.1, 07/2017, Available at: www.etsi.org/deliver/etsi_gs/MEC/001_099/012/01.01.01_60/gs_MEC012v010101p.pdf.
19. ETSI GS MEC 013, "Mobile Edge Computing (MEC); Location API," v. 1.1.1, 07/2017, Available at: www.etsi.org/deliver/etsi_gs/MEC/001_099/013/01.01.01_60/gs_MEC013v010101p.pdf.
20. OMA-TS-REST-NetAPI-ZonalPresence-V1-0-20160308-C, "RESTful Network API for Zonal Presence."
21. OMA-TS-REST-NetAPI-ACR-V1-0-20151201-C, "RESTful Network API for Anonymous Customer Reference Management."
22. SCF084, "Small Cell Zone Services: RESTfule Bindings." Available at: http://scf.io/en/documents/084_-_Small_Cell_Zone_services_RESTful_Bindings.php.
23. SCF 152, "Small Cell Services API." Available at: http://scf.io/en/documents/152_-_Small_cell_services_API.php.
24. ETSI GS MEC 014, "Mobile Edge Computing (MEC); UE Identity API," v. 1.1.1, 02/2018. Available at: www.etsi.org/deliver/etsi_gs/MEC/001_099/014/01.01.01_60/gs_MEC014v010101p.pdf.
25. ETSI, "MEC in an Enterprise Setting: A Solution Outline," 2018. Available at: www.etsi.org/images/files/ETSIWhitePapers/etsi_wp30_MEC_Enterprise_FINAL.pdf.
26. ETSI GS MEC 015, "Mobile Edge Computing (MEC); Bandwidth Management API," v. 1.1.1, 07/2017. Available at: www.etsi.org/deliver/etsi_gs/MEC/001_099/015/01.01.01_60/gs_MEC015v010101p.pdf.

27. OpenStack, "Cloud Edge Computing: Beyond the Data Center." Available at: www.openstack.org/assets/edge/OpenStack-EdgeWhitepaper-v3-online.pdf (accessed Jan. 2019).
28. VmWare Project Dimension, www.vmware.com/products/project-dimension.html (accessed Jan. 2019).
29. ETSI GS MEC 010-1, "Mobile Edge Computing (MEC); Mobile Edge Management; Part 1: System, host and platform management," v. 1.1.1, 10/2017. Available at: www.etsi.org/deliver/etsi_gs/MEC/001_099/01001/01.01.01_60/gs_MEC01001v010101p.pdf.
30. ETSI GS MEC 010-2, "Mobile Edge Computing (MEC); Mobile Edge Management; Part 2: Application Lifecycle, Rules and Requirements Management," 07/2017. Available at: www.etsi.org/deliver/etsi_gs/MEC/001_099/01002/01.01.01_60/gs_MEC01002v010101p.pdf.
31. ETSI GS MEC 016, "Mobile Edge Computing (MEC); UE Application Interface," v. 1.1.1, 09/2017. Available at: www.etsi.org/deliver/etsi_gs/MEC/001_099/016/01.01.01_60/gs_MEC016v010101p.pdf.
32. ETSI GR MEC 017, "Mobile Edge Computing (MEC); Deployment of Mobile Edge Computing in an NFV Environment," v. 1.1.1, 02/2018. Available at: www.etsi.org/deliver/etsi_gr/MEC/001_099/017/01.01.01_60/gr_MEC017v010101p.pdf.
33. ETSI GS MEC-IEG 004, "Mobile-Edge Computing (MEC); Service Scenarios," v. 1.1.1, 11/2015. Available at: www.etsi.org/deliver/etsi_gs/MEC-IEG/001_099/004/01.01.01_60/gs_MEC-IEG004v010101p.pdf.
34. ETSI GS MEC 002, "Multi-access Edge Computing (MEC); Phase 2: Use Cases and Requirements," v. 2.1.1, 10/2018. Available at: www.etsi.org/deliver/etsi_gs/MEC/001_099/002/02.01.01_60/gs_MEC002v020101p.pdf.
35. Hewlett Packard Enterprise, "Edge Video Analytics. HPE Edgeline IoT Systems with IDOL Media Server Enable Exceptional Scene Analysis & Object Recognition Performance." Available at: https://support.hpe.com/hpsc/doc/public/display?docId=emr_na-c05336736&docLocale=en_US.
36. HPE, AWS, Saguna, "A Platform for Mobile Edge Computing." Available at: www.saguna.net/blog/aws-hpe-saguna-white-paper-platform-for-mobile-edge-computing/ (accessed Jan. 2019).
37. 3GPP TR 22.866, "Study on Enhancement of 3GPP Support for 5G V2X Services (Release 16)," v. 16.2.0, 12/2018.
38. ETSI GS MEC 022, "Multi-access Edge Computing (MEC); Study on MEC Support for V2X Use Cases," v. 2.1.1, 09/2018.
39. ETSI, "Cloud RAN and MEC: A Perfect Pairing," 2018. Available at: www.etsi.org/images/files/ETSIWhitePapers/etsi_wp23_MEC_and_CRAN_ed1_FINAL.pdf.
40. ETSI, "MEC Deployments in 4G and Evolution Towards 5G," 2018. Available at: www.etsi.org/images/files/ETSIWhitePapers/etsi_wp24_MEC_deployment_in_4G_5G_FINAL.pdf.
41. ETSI, "MEC in 5G Networks," 2018. Available at: www.etsi.org/images/files/ETSIWhitePapers/etsi_wp28_mec_in_5G_FINAL.pdf.

42. 3GPP TS 23.501 V15.3.0, "3rd Generation Partnership Project; Technical Specification Group Services and System Aspects; System Architecture for the 5G System; Stage 2 (Release 15)".
43. 3GPP TS 29.500 V15.1.0, "3rd Generation Partnership Project; Technical Specification Group Core Networks and Terminals; 5G System; Technical Realization of Service Based Architecture; Stage 3 (Release 15)".
44. STL Partners, HPE, Intel, "Edge Computing, 5 Viable Telco Business Models," 11/2017. Available at: https://h20195.www2.hpe.com/v2/get-pdf.aspx/a00029956enw.pdf (accessed Jan. 2019).
45. "Report ITU-R M.2370-0, IMT Traffic Estimates for the Years 2020 to 2030," 07/2015. Available at: www.itu.int/dms_pub/itu-r/opb/rep/r-rep-m.2370-2015-pdf-e.pdf.
46. "Cisco Visual Networking Index: Global Mobile Data Traffic Forecast Update, 2016–2021 White Paper," 02/2017. Available at: www.cisco.com/c/en/us/solutions/collateral/service-provider/visual-networking-index-vni/mobile-white-paper-c11-520862.html.
47. "Cisco Global Cloud Index: Forecast and Methodology, 2016–2021 White Paper," 11/2018. Available at: www.cisco.com/c/en/us/solutions/collateral/service-provider/global-cloud-index-gci/white-paper-c11-738085.html.
48. "Ericsson Mobility Report," 11/2018. Available at: www.ericsson.com/assets/local/mobility-report/documents/2018/ericsson-mobility-report-november-2018.pdf.
49. "Brian Krzanich, Talk at InterDrone," Las Vegas, NV, September 6–8 2017. Available at: https://uavcoach.com/intel-ceo-interdrone/.
50. Recommendation ITU-R M.2083, "IMT Vision – Framework and Overall Objectives of the Future Development of IMT for 2020 and Beyond," 09/2015.
51. ITU WP 5D, Draft new Report ITU-R M. [IMT-2020.TECH PERF REQ] – "Minimum Requirements Related to Technical Performance for IMT-2020 Radio Interface(s)," 02/2017.
52. B. Barani, "Network Technologies," European Commission, Panel 5G: From Research to Standardisation, Austin, 8 December 2014.Available at: www.irisa.fr/dionysos/pages_perso/ksentini/R2S/pres/Bernard-EC-Panel-R2S-2014.pdf.
53. GSMA, "5G Spectrum – GSMA Public Policy Position," 11/2018. Available at: www.gsma.com/spectrum/wp-content/uploads/2018/11/5G-Spectrum-Positions.pdf.
54. A. Manzalini, "Multi-access Edge Computing: Decoupling IaaS-PaaS for Enabling New Global Ecosystems," Berlin, 19–20 September 2018. Edge Computing Congress, 2018.
55. ETSI ISG MEC, Available at: www.etsi.org/technologies/multi-access-edge-computing.
56. ETSI White Paper, "Mobile Edge Computing: A Key Technology Towards 5G," 09/2015. Available at: www.etsi.org/images/files/ETSIWhitePapers/etsi_wp11_mec_a_key_technology_towards_5g.pdf.
57. ETSI ISG NFV, Available at: www.etsi.org/technologies/nfv

58. ETSI GS MEC 002 V1.1.1 (2016-03), "Mobile Edge Computing (MEC); Technical Requirements." Available at: www.etsi.org/deliver/etsi_gs/MEC/001_099/002/01.01.01_60/gs_MEC002v010101p.pdf.
59. ETSI GS MEC 003 V1.1.1 (2016-03), "Mobile Edge Computing (MEC); Framework and Reference Architecture." Available at: www.etsi.org/de-liver/etsi_gs/MEC/001_099/003/01.01.01_60/gs_MEC003v010101p.pdf.
60. ETSI MEC 017, "MEC in NFV Deployment." Available at: www.etsi.org/de-liver/etsi_gr/MEC/001_099/017/01.01.01_60/gr_mec017v010101p.pdf.
61. ETSI GS MEC 003 V2.1.1 (2019-01), "Multi-access Edge Computing (MEC); Framework and Reference Architecture." Available at: www.etsi.org/deliver/etsi_gs/MEC/001_099/003/02.01.01_60/gs_mec003v020101p.pdf.
62. 3GPP SA2, "Study on Enhancement of Support for Edge Computing in 5GC," S2–1902838- Study Item Description, 03/2019. Available at: www.3gpp.org.
63. 3GPP SA5, "Study on Management Aspects of Edge Computing," xx. Available at: www.3gpp.org/ftp/SPecs/archive/28_series/28.803/.
64. 3GPP SA6. "Study on Application Architecture for Enabling Edge Applications," S6–190238- Study Item Description, 01/2019. Available at: www.3gpp.org.
65. 5GAA White Paper, "Toward Fully Connected Vehicles: Edge Computing for Advanced Automotive Communications," 12/2017. Available at: http://5gaa.org/wp-content/uploads/2017/12/5GAA_T-170219-whitepaper-EdgeComputing_5GAA.pdf.
66. ETSI White Paper, "MEC Deployments in 4G and Evolution Towards 5G," 02/2018. Available at: www.etsi.org/images/files/ETSIWhitePapers/etsi_wp24_MEC_deployment_in_4G_5G_FINAL.pdf.
67. ETSI White Paper, "MEC in 5G Networks," 06/2018. Available at: www.etsi.org/images/files/ETSIWhitePapers/etsi_wp28_mec_in_5G_FINAL.pdf.
68. ETSI ISG MEC, DGR/MEC-0031 Work Item, "MEC 5G Integration." Available at: https://portal.etsi.org/webapp/workProgram/Report_Schedule.asp?WKI_ID=56729.
69. NGMN Alliance, "Description of Network Slicing Concept," 01/2016.
70. 3GPP TS 23.501 V15.4.0 (2018-12), "System Architecture for the 5G System." Available at: www.3gpp.org/ftp/specs/archive/23_series/23.501/.
71. 3GPP TS 22.261 V16.6.0 (2018-12), "Service Requirements for the 5G System." Available at: www.3gpp.org/ftp/specs/archive/22_series/22.261/.
72. 3GPP TS 28.531 V16.0.0 (2018-12), "Management and Orchestration; Provisioning." Available at: www.3gpp.org/ftp/Specs/Archive/28_series/28.531/.
73. 3GPP TS 28.532 V15.1.0 (2018-12), "Management and Orchestration; Generic Management Services." Available at: www.3gpp.org/FTP/Specs/archive/28_series/28.532/.

74. DGR/MEC-0024 Work Item, "MEC Support for Network Slicing." Available at: https://portal.etsi.org/webapp/WorkProgram/Report_WorkItem.asp?wki_id=53580.
75. GSMA Network Slicing Task Force, Available at: www.gsma.com/futurenetworks/technology/understanding-5g/network-slicing/.
76. STL Partners, HPE, Intel, "Edge Computing, 5 Viable Telco Business Models," 11/2017. Available at: https://h20195.www2.hpe.com/v2/getpdf.aspx/a00029956enw.pdf (accessed Jan. 2019).
77. "Chetan Sharma consulting, Operator's Dilemma (And Opportunity): The 4th Wave; White Paper," 2012. Available at: www.chetansharma.com/publications/operators-dilemma-and-opportunity-the-4th-wave/.
78. "Digital TV Research, Global OTT Revenue Forecasts," 11/2018. Available at: www.digitaltvresearch.com/ugc/press/245.pdf.
79. GSMA, Network 2020 Programme, "Unlocking Commercial Opportunities – From 4G Evolution to 5G," 02/2016. Available at: www.gsma.com/futurenetworks/wp-content/uploads/2016/02/704_GSMA_unlocking_comm_opp_report_v5.pdf.
80. 5G-ACIA, "5G Non-Public Networks for Industrial Scenarios," White Paper, 03/2019. Available at: www.5g-acia.org/index.php?id=6958.
81. Riot Research, "Connected Car Revenue Forecast 2017–2023," 02/2018. Available at: https://rethinkresearch.biz/wp-content/uploads/2018/02/Connected-Car-Forecast.pdf.
82. Cisco, "Cisco Visual Networking Index: Forecast and Trends, 2017–2022 White Paper." Available at: www.cisco.com/c/en/us/solutions/collateral/service-provider/visual-networking-index-vni/white-paper-c11-741490.html (accessed Feb. 2018).
83. Allied Market Research, "Global Factory Automation Market Overview," 06/2018. Available at: www.alliedmarketresearch.com/factory-automation-market.
84. Zion Market Research, "Global Smart Manufacturing," 05/2018. Available at: www.zionmarketresearch.com/report/smart-manufacturing-market.
85. Markets and Markets, "eHealth Market," 03/2018. Available at: www.marketsandmarkets.com/Market-Reports/ehealth-market-11513143.html?gclid=EAIaIQobChMIsO3k5eG-4QIVj-F3Ch0v5Qf2EAAYASAAEgIDIfD_BwE.
86. IoT Analytics, "State of the IoT 2018: Number of IoT Devices Now at 7B–Market Accelerating," 08/2018. Available at: https://iot-analytics.com/state-of-the-iot-update-q1-q2-2018-number-of-iot-devices-now–7b/.
87. iGR White Paper, "The Business Case for MEC in Retail: A TCO Analysis and Its Implications in the 5G Era," 07/2017. Available at: https://networkbuilders.intel.com/blog/opportunities-for-multi-access-edge-computing-and-5g-in-retail-total-cost-of-ownership-analysis.
88. Ericsson Technology Review, "Distributed Cloud – A Key Enabler of Automotive and Industry 4.0 Use Cases," 11/2018. Available at: www.ericsson.com/en/ericsson-technology-review/archive/2018/distributed-cloud.

89. "Cisco Global Cloud Index: Forecast and Methodology, 2016–2021 White Paper," 11/2018. Available at: www.cisco.com/c/en/us/solutions/collateral/service-provider/global-cloud-index-gci/white-paper-c11-738085.html.
90. Intel White Paper, "Data Center Strategy Leading Intel's Business Transformation," 11/2017. Available at: www.intel.com/content/dam/www/public/us/en/documents/white-papers/data-center-strategy-paper.pdf.
91. D. Sabella, "5G Experimental Activities in Flex5Gware Project – Focus on RAN Virtualization," IEEE 5G Berlin Summit, Fraunhofer-Forum Berlin, 2 November 2016. Available at: www.5gsummit.org/berlin/docs/slides/Dario-Sabella.pdf.
92. D. Thanh, I. Jørstad, "The Mobile Phone: Its Evolution from a Communication Device to a Universal Companion," 2005. Available at: www.researchgate.net/publication/239608908_The_mobile_phone_Its_evolution_from_a_communication_device_to_a_universal_companion.
93. N. Islam, R. Want, "Smartphones: Past, Present, and Future," *IEEE Pervasive Computing*, vol. 13, issue 4, October–December 2014. Available at: https://ieeexplore.ieee.org/document/6926722.
94. P. Rubin, "Wired, Eye Tracking Is Coming to VR Sooner than You Think. What Now?" 23 March 2018. Available at: www.wired.com/story/eye-tracking-vr/.
95. R. Berger, "Digital Factories: The Renaissance of the U.S. Automotive Industry." Available at: www.rolandberger.com/en/Publications/pub_digital_factories.html.
96. Ericsson, Technology Trends 2018, "Five Technology Trends Augmenting the Connected Society," 2018. Available at: www.ericsson.com/en/ericsson-technology-review/archive/2018/technologytrends-2018.
97. ETSI GS MEC 002 V2.1.1 (2018-10), "Multi-access Edge Computing (MEC); Phase 2: Use Cases and Requirements." Available at: www.etsi.org/deliver/etsi_gs/MEC/001_099/002/02.01.01_60/gs_MEC002v020101p.pdf.
98. Intel White Paper, "Intel IT's Data Center Strategy for Business," 2014. Available at: www.insight.com/content/dam/insight-web/en_US/article-images/whitepapers/partner-whitepapers/intel-it-s-data-center-strategy-for-business-transformation.pdf.
99. EARTH Project Deliverable D2.1, "Economic and Ecological Impact of ICT." Available at: https://bscw.ict-earth.eu/pub/bscw.cgi/d38532/EARTH_WP2_D2.1_v2.pdf
100. G. Auer et al., "Cellular Energy Efficiency Evaluation Framework," Vehicular Technology Conference (VTC Spring), 2011 IEEE 73rd; 15–18 May 2011; Budapest, Hungary, 2011. ISSN: 1550–2252. Available at: http://ieeexplore.ieee.org/xpl/freeabs_all.jsp?arnumber=5956750.

101. D. Sabella et al., "Energy Efficiency Benefits of RAN-as-a-Service Concept for a Cloud-Based 5G Mobile Network Infrastructure," *IEEE Access*, vol. 2, December 2014; ISSN: 2169–3536, doi:10.1109/ACCESS.2014.2381215.
102. China Mobile Research Institute, "C-RAN The Road Towards Green RAN," White Paper, Version 2.5, 10/2011. Available at: https://pdfs.semanticscholar.org/eaa3/ca62c9d5653e4f2318aed9ddb8992a505d3c.pdf.
103. D. Rapone, D. Sabella, M. Fodrini, "Energy Efficiency Solutions for the Mobile Network Evolution Toward 5G: An Operator Perspective," The Fourth IFIP Conference on Sustainable Internet and ICT for Sustainability, SustainIT2015, Madrid, 15 April 2015. Available at: www.networks.imdea.org/sustainit2015/t-program.html.
104. D. Sabella et al., "Energy Management in Mobile Networks Towards 5G," In: M. Z. Shakir, M. A. Imran, K. A. Qaraqe, M.-S. Alouini, A. V. Vasilakos, (eds.), *Energy Management in Wireless Cellular and Ad-hoc Networks*, Volume 50 of the series Studies in Systems, Decision and Control, Springer, Switzerland, 2015, pp. 397–427.
105. ETSI GS MEC-IEG 005 V1.1.1 (2015-08), "Mobile-Edge Computing (MEC); Proof of Concept Framework." Available at: www.etsi.org/deliver/etsi_gs/MEC-IEG/001_099/005/01.01.01_60/gs_MEC-IEG005v010101p.pdf.
106. Akraino Edge Stack website, www.lfedge.org/projects/akraino/.
107. OSF Edge Computing Group website, www.openstack.org/edge-computing/.
108. 5G Infrastructure Public Private Partnership (5G PPP) website, https://5g-ppp.eu/.
109. 5G Automotive Association (5GAA) website, https://5gaa.org/.
110. ETSI ISG MEC, DGS/MEC-0030V2XAPI' Work Item "Mobile-Edge Computing (MEC); MEC V2X API" Available at: https://portal.etsi.org/webapp/WorkProgram/Report_WorkItem.asp?wki_id=54416.
111. L. Baltar, M. Mueck, D. Sabella, "Heterogeneous Vehicular Communications – Multi-Standard Solutions to Enable Interoperability", 2018 IEEE Conference on Standards for Communications and Networking (CSCN), Paris (France), 29-31 Oct. 2018. Available at: https://ieeexplore.ieee.org/document/8581726.

Index

Q

R

S